THE
NEUROSCIENCE
OF
YOU

THE
NEUROSCIENCE
OF
YOU

How Every Brain Is Different and
How to Understand Yours

CHANTEL PRAT

DUTTON

DUTTON

An imprint of Penguin Random House LLC
penguinrandomhouse.com

Illustrations on pages 1, 22, and 82 by Andrea Stocco.
Illustration on page 17 modified from Elisa Galliano's neuron entry in SciDraw.io, https://doi.org/10.5281/zenodo.3926535.
Image on page 296 reproduced with permission from Baron-Cohen et al., 2001.
All other images courtesy of the author.

LIBRARY OF CONGRESS CATALOGING-IN-PUBLICATION DATA
has been applied for.

ISBN 9781524746605 (hardcover)
ISBN 9781524746612 (ebook)

Printed in the United States of America
1 3 5 7 9 10 8 6 4 2

Book design by Nancy Resnick

To Jasmine, Andrea, and Coccolina—for loving me as I am

CONTENTS

CONTENTS

FROM MY BRAIN TO YOURS

They say that everyone has a book in them, but no one ever tells you how hard it is to get that book *out* of you. Well, they didn't tell *me* anyway. To be fair, I probably wouldn't have listened. As it turns out, my brain is more of a "touch the stove to see *how* hot" kind of learner. To be honest, I'm thankful for it. Because even if I get burned now and then along the way, if I had a "because they said so" type of brain, I wouldn't have done most of the hard things that prepared me to write this book in the first place. And if you learn *half* as much about your brain when you read it as I did when I wrote it, it will definitely have all been worth it.

Suffice to say that my first book-writing experience has been anything but "normal," if there is such a thing. A big part of it involved the experiment we all participated in that began in 2020—and I'm pretty sure none of us signed a consent form. You know, the one centered on a virus? I'd like to think of it as a radical exploration of what psychologists have called the nature-versus-nurture question: How much of what makes you *you* is inherent in your biological makeup, and how much is a response to your environment? When the COVID-19 pandemic hit, many of us traded the routine parts of our daily lives for pervasive anxiety about our health and the safety of our loved ones.

Fortunately, my "day job" as a scientist and professor at the University of Washington in Seattle gave me some tools for understanding what might happen to me under these circumstances. But for reasons you'll read about in the second half of this book, my knowing better didn't immediately translate into my doing better. Instead, I watched my life transform with equal parts fascination and horror. I was captivated by the differences between how I felt and how the people around me seemed to be coping with the changes in their routines. Some of them got into "the best shape of their lives," while I remained stagnant. Others exchanged recipes and became obsessed with baking the perfect loaf of sourdough bread. Not only did I cook *less than ever*, I didn't do *any* of the things I always said I would do if I had more time.

Instead, I tried my best to finish Netflix. I cajoled my husband into playing dozens of hours of Pandemic, a board game in which you try to save the world from a virus outbreak. I ate like crap. I drank more than normal. And in the moments of stillness, while gazing at my increasingly protruding navel, I found myself asking the very question that got me into this field in the first place.

"Why am I *like* this?"

The answer is pragmatically simple but biologically and philosophically complicated enough to fill a whole shelfful of books.

My *brain* makes me this way.

I remember the *exact* moment I first had this realization, and how swiftly it changed my life forever. I was nineteen years old and, after watching one-too-many episodes of *Doogie Howser, M.D.,* was on my way to applying to med school. To meet my last requirement, I signed up for a psychology course at the local junior college that didn't interfere with my day job, selling shoes at Kinney's in the mall.

And during our first class, the instructor described the story of Phineas Gage.

Gage was a railway worker who made a mistake in 1848 that caused an iron spike to get blasted through his left cheek and out the top of his head. When it did, it took a decent-size chunk of his brain with it. Surviving such an injury would be remarkable even with today's medical practices—so the fact that Gage got up and literally walked away from the accident is incredible in and of itself. Eventually, many of his physical and mental abilities returned to "normal." But the damage Gage sustained to his frontal lobe left his *personality* fundamentally, and permanently, changed. While Gage was once a well-respected and dependable man, capable of forming and executing rational plans, his physician described him as "fitful, irreverent . . . manifesting but little deference for his fellows, impatient of restraint or advice when it conflicts with his desires, at times pertinaciously obstinate, yet capricious and vacillating, devising many plans of future operations, which are no sooner arranged than they are abandoned in turn for others appearing more feasible." To put it simply, Gage was not the same *person* after his brain injury.

This *fascinated* me.

I left class trying to wrap my mind around the fact that the human brain is an organ, just like the heart or the lungs, but that the functioning of this organ makes you *you*. The lungs oxygenate the blood. The heart circulates the oxygenated blood through the body, and then your brain uses that oxygenated blood to create the energy that gives rise to every thought, feeling, emotion, and action that you identify as your own. Change the brain and you change the person.

What I realized about three months into the pandemic is that on a smaller (and, hopefully, less permanent) scale, *my* brain was changing. Soaked in cortisol—a neurochemical related to prolonged stress—my brain was struggling to find a balance between the "should do" and "want to do" urges. And I don't know who needs to hear this, but being stressed also *majorly* kills creativity.

Thankfully, while writing the "Mixology" chapter, I had the aha moment that gave me some much-needed perspective. Among other things, it reminded me *why* people were responding to the pandemic in different ways. At the end of the day, people respond to stress differently for the same reason that some people feel paranoid when they smoke weed for the first time, while others just feel hungry. It all goes back to the nature-versus-nurture question—and the answer is almost always a combination of both. Baseline differences in our biology *combine* with our lived experiences to shape the way we think, feel, and *respond* to our environmental changes. And I know that *my brain* did the best it could under the circumstances. It always does. I sincerely hope *your brain* will enjoy learning about itself through the fruits of our labor.

THE
NEUROSCIENCE
OF
YOU

THE NEUROSCIENCE OF YOU 101

Let me start by saying how excited I am for the opportunity to introduce *you* to *your* brain! It doesn't seem right, after all, that I might know more about the thing that's driving you around this world than you do. To be fair, I've been at this for a while now, so I've got a bit of a head start. Since I got my first job in a brain development lab in the mid-'90s, I've been working at the intersection of neuroscience, psychology, linguistics, and neural engineering. The goal of my research is straightforward, but not simple—to figure out how *differences* in brain functioning shape the way people process information. In short, I want to understand what makes people like you tick.

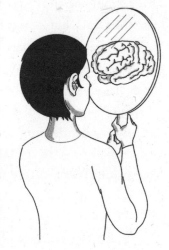

Though I'm sure most of you understand, on some level, that your unique ways of thinking, feeling, and behaving are related to the way your brain works, the vast majority of neuroscience books on the shelf adopt the one-size-fits-all approach that has dominated the field for more than a century. And let's face it: Things that are one-size-fits-all don't fit *anyone* very well. In fact, what I have learned in my professional life echoes what I have observed in my most interesting and rewarding interactions with people in the real world.

We *do not* all work the same.

This book goes beyond describing how *most* brains work, to allow you to get a better understanding of how *your* brain works. Because, at risk of sounding cliché, every brain really is unique. Even identical twins conjoined at birth have different brains! And though this might come as a bit of a surprise, a number of the differences between healthy human brains can have profound effects on the way they work.

Remember "The Dress" that took the Internet by storm in 2015 because people couldn't agree about whether it was blue and black or white and gold?[1] I have to assume that the reason *millions* of people were captivated by that picture is because the version of reality that our brains create for us is *so convincing*. It's a bit shocking to learn that something as elementary as the color of a dress might be open to interpretation. But by the time you've finished the "Adapt" chapter, the science behind how different brains understand the color of the Dress should be clear to you. This won't change the way you experience the Dress, but it might give you a new perspective on the whole "seeing is believing" adage. Because, as you'll soon learn, differences in how our brains work shape not only the way we see the world but the decisions we make about how to behave in it.

Ready to get started learning about yours?

Put your hands into fists and turn them so both thumbs are facing you. Now bring the knuckles of your fingers together and voilà! You've got a pretty decent model of *the size* of your brain.

---

1. If you managed to miss this phenomenon, Wikipedia has a nice entry titled "The Dress" that includes the original picture.

It's kind of humbling, isn't it?

Though it might be *smaller* than you imagined, it is *mighty*. That three(ish)-pound bundle of 86 billion (give or take) signal-producing brain cells called neurons is single-handedly[2] responsible for translating physical energy in the world "out there" into *your* version of reality. Of course, it also controls most of your bodily functions and keeps you alive in its "spare time." To accomplish its important jobs, your brain uses *at least* 20 percent of your body's energy resources at any given time, even though it only makes up about 2 percent of its total weight. In other words, your brain is *expensive*.

And don't even get me started about how brilliantly engineered it is. To get the most brainpower into a head you can still manage to carry around, evolutionary pressure caused "gyrification" of big brains—a process by which their surfaces folded in on themselves to fit in smaller spaces. It's kind of like taking a piece of paper and crumpling it into a ball. If you were to "uncrumple" the computationally powerful layer of neurons that covers your brain—the cerebral cortex—it would cover about the same surface area as two medium pizzas.[3] And because your brain cells are jammed in there so tightly, your brain doesn't even have room to store backup fuel reserves, like most of your other organs do. As a result, it requires a *constant* supply of glucose, even when you're sleeping. Suffice to say that if we haven't "maxed out" when it comes to the amount of brainpower our bodies can support, we're damn close.

But you might still be wondering what the size of your fists can tell you about how your brain works. I should probably warn you up front—if you're hoping to read this book and learn that because you have really big hands, your brain is better, faster, and stronger than

---

2. Well, I guess it's double-handedly in this case—Sorry, that was a total dad joke. My brain made me do it! If you decide to quit reading the footnotes right now, it's OK. I deserve it.

3. The actual size is nearly 2.5 square feet.

average, you are probably going to be disappointed.[4] Don't get me wrong, bigger *is* better in some circumstances, but that's not what this book is about. Most of the important characteristics of the organ that makes you *you* are more nuanced than that.

Take, for example, the research paper titled "Big-Brained People Are Smarter" by Michael McDaniel. In it, McDaniel analyzed the relation between brain volume and performance on standardized intelligence tests using data collected from over 1,500 people. As you might suspect from the spoiler in the title, people with bigger brains *do* tend to score higher on intelligence tests.[5] According to his analysis, the *correlation* between the two variables—a statistic that estimates how much the value of one variable (like brain size) can tell you about the value of another one (like intelligence scores) was .33. If you square this number, and multiply it by 100, it gives you something more interpretable—the percentage of variability in one value explained by the other. In this case, that value is 10.89 percent, which means that if you're trying to explain what makes people perform differently on intelligence tests, knowing how big their brains are will get you almost 11 percent of the way there. While that's clearly a decent-size piece of the puzzle, I *hope* it makes you wonder what explains the *other* 89 percent, especially since your brain is 100 percent responsible for your performance on any test.

---

4. Well, statistically speaking, if you have big hands, you might be stronger than average, but *that* is beside the point.

5. I try to be intentional about not using words like "smarter" or "more intelligent" to describe people who score higher on intelligence tests. There is still plenty of debate within the scientific community about what intelligence is and how to measure it. I tend to side with Edwin Boring, who wrote in 1923 that "intelligence is what intelligence tests test."

How your brain is engineered

The truth about differences between human brains—or at least the version of it that my brain has created for me—is more complicated than just bigger being better.[6] This makes sense, if you consider the fact that evolution has *already* been hard at work for hundreds of millions of years cramming more and more horsepower into our heads. But the evolutionary pressures that shaped *your* brain didn't care specifically about how big it got. Instead, the *success* of a brain is measured by its ability to drive the body it inhabits around the world in a way that allows it to live long enough to find another brain that's willing to reproduce with it. Over time, many different types of brains evolved, each optimally engineered to pilot different kinds of bodies through the specific environments they inhabit.[7]

I should also warn you that this book has little to do with how to find the right partner, though the last chapter of the book—"Connect"—does describe the challenges two brains encounter when they're trying to communicate through the different worlds they've created for themselves. Instead, we're going to focus on the massive information-processing engine that is *your brain*, and yours alone. Much like the parts of a vehicle's engine translate energy from batteries or fuel combustion into mechanical forces that move it through the world, the goal of any brain is to translate physical energy from the environment it inhabits into information it can use to make the decisions that drive it through the world in a way that maximizes its success.

But here's the catch—the universe your brain operates in is

---

6. If it weren't, sperm whales with their massive, twenty-pound brains would run the world.

7. The octopus, with eight big brains to control each of its arms, and a small central nervous system to coordinate activities between them, is a striking example of this. I suspect that if you asked an octopus brain to pilot a human body, it would have a hell of a time putting on its pants.

essentially *infinite* and continuously changing. Your brain, mighty though it may be, is finite. It has to process the world "out there" in discrete, bite-size chunks. It's kind of like taking a series of low-resolution snapshots and then building a movie out of them. Doing so requires a million decisions about which bits of information are the most important, and how to "connect the dots" when there are missing pieces. As you'll read in this book, individual brains have different ways of trying to manage their inherent limitations.

Much like engines have different mechanisms for translating energy into motion (for example, the number of cylinders or type of transmission they have), every brain has a set of design features that shapes the way it reconstructs the incomplete data it takes in, and generates the thoughts, feelings, and decision-making patterns that drive *you*. And this is how we're going to go about the business of figuring out how *your* brain works. Because obviously, without all the fancy equipment I use in the lab to measure it more directly, the best we can do is to reverse-engineer your brain, based on the way you think, feel, and act.

In the chapters that follow, I've included a series of assessments to help you get a better idea of how your brain is designed.[8] As we get into the process of reverse-engineering it, you'll start to learn about the costs and benefits associated with each of the different design features we discuss. This makes sense, if you consider how long evolution has been working to weed out brain designs that don't work well for anyone in any situation. Sure, when they are faced with a specific kind of problem, one type of brain might do better than another. But there is almost always another situation in which the second brain type will excel.

In other words, trying to decide what type of brain design is "the best" is kind of like trying to decide whether a Honda Civic is better

---

8. If your find yourself wanting to learn more about your brain than I can fit on the page, feel free to visit the "Research" tab on my website, chantelprat.com, for links to more brain assessments.

or worse than a Subaru Outback. Sure, I have a personal opinion about this, but in truth, they are two different cars that have been engineered to meet different needs. Deciding which one is better depends a lot on what you need your car to do. I *hope* you can keep this in mind when you start figuring out how your brain works. This book is more about "finding your lane" than "winning a race" with it!

The brains of London taxi drivers, which became temporarily famous in the year 2000, perfectly illustrate this idea. Before getting licensed to drive a cab in London, a person has to pass an incredibly difficult test with an equally intimidating name—"The Knowledge." The test involves memorizing the layout of more than 20,000 streets in the Greater London area—a feat that involves an incredible amount of memory resources. As you might suspect, this pretty quickly weeds out candidates like me, whose brains are all RAM and no hard drive, so to speak. In fact, fewer than 50 percent of the people who sign up for taxi driver training pass the test, even after spending two or three years studying for it! And as it turns out, the brains of London cabbies are different from non-cab-driving humans in ways that reflect their herculean memory efforts. In fact, the part of the brain that has been most frequently associated with spatial memory, the tail of the sea horse–shaped brain region called the hippocampus, is *bigger* than average in these taxi drivers.[9] But here's a fun fact that didn't make it into their first fifteen minutes of fame—the head of the hippocampus is *smaller* than average in the taxi drivers!

To figure out the implications of this particular brain design, Eleanor Maguire, the Irish neuroscientist who discovered the remarkable brains of these drivers, conducted a follow-up study. To control for the environmental demands that the taxi drivers' brains operate in, moving a vehicle through busy streets without bumping into

---

9. If you're interested in how it got that way, buckle your seatbelt. We're going to talk about that in the next section.

things, she compared their memory performance to another group whose brains operate in a similar environment—London bus drivers. The results of this literal head-to-head comparison were pretty fascinating. While the taxi drivers outperformed the bus drivers on tests that involved recognizing landmarks in London, or judging the distance between familiar places in the city, the bus drivers outperformed taxi drivers on tests that involved drawing complex figures from memory, or remembering lists of words. In other words, the brains of the taxi drivers showed a specific type of memory enhancement—one that allowed them to acquire a massive amount of spatial information from the maps they studied. However, that enhancement also seemed to come with a measurable *cost* to other memory functions, as it crowded out nearby brain regions with other jobs to perform. And though I'm sure you could engage a group of cabbies or bus drivers in a lively debate about which group is *smarter*, both performed equally well on more tests than they differed on, including things that are central to many environments, like the ability to remember stories or recognize people's faces.

The example of the two types of drivers' brains nicely illustrates many of the principles of brain engineering that this book is organized around. The first is the notion of costs and benefits. If Maguire had not been motivated to understand the whole story, it would have been very easy to decide that bigger *was* better. Taxi drivers have bigger spatial memory regions and are better able to memorize a massive number of maps. And if I were to ask the average person off the street whether they'd like to have a better memory, most of them would say yes. But what if I were to ask you whether you'd rather be able to memorize massive amounts of spatial information, or to memorize your grocery list or sketch something you've seen once from memory? Then the answer would probably depend on what you need, or like, to do with your brain, right?

And this brings me back to my second point about brain designs.

It doesn't make sense to decide that one is better than the other without thinking about what you need to *do* with it. Unlike the Honda Civic versus Subaru Outback comparison, your brain is also engineered by the environments it gets put in and the tasks you ask it to accomplish. In other words, your brain might currently be a Subaru Outback, a Honda Civic, or even a Ford F-150, but you were born closer to a Volkswagen Beetle, or a Fiat 500, and your experiences *helped* to shape you into what you are today.

In the pages that follow, I plan to help you understand your brain better by describing some of the design features that are most influential in driving you around the environment you inhabit. Starting from the inside out, in Part 1, we'll discuss some of the biological pressures that shape brains in different ways, from the asymmetries that give rise to specialized brain functions to the chemicals that fuel your brain's communication systems. Then, in Part 2, we'll take a look at how external pressures both shape your brain and interact with its intrinsic design features. What are the jobs a brain needs to do to be successful, and how might different ways of getting these jobs done be reflected in different types of brain designs? From the need to adapt to different environments to the desire to understand and connect with others, some of our most remarkable differences show up when we watch brains responding to the variety of situations we ask them to drive us through. But before we get started discussing how all of this works, I'd like to give you a bit more of the theoretical background, to help you to understand what it *means* when I say "Your brain makes you that way."

What does it mean to be different?

I'll be the first one to admit how much comfort I find in a good book, fiction or nonfiction, that helps me feel like some of the things that I think are really strange about myself are actually pretty *normal*. But my understanding of what makes something *normal* or *abnormal* is probably different from yours, so this seems like a good place to start our discussion. The first thing to note is that the distinction between "normal" and "abnormal" is almost never a binary one. It's not like those of us *in the know* look at a group of people through our scientific lenses and think "normal, normal, normal, WEIRD, normal, normal." It just doesn't work like that.

Instead, whether you're studying someone's general level of optimism about the future or their brain size, there is almost always a range of values that describe the characteristic you're interested in. The question then becomes—Are you "within the normal range" or "outside" of it? But how do we decide where the boundary is?

And here's something not everyone understands—you can't *scientifically* define what's "normal" or "abnormal" without understanding the nature of how people differ. When we do, we need to hold two different ways of defining *normal* in mind: (1) How typical or atypical is a particular way of being? and (2) How functional or dysfunctional is it?

Let's take attention deficit/hyperactivity disorder (ADHD) as an example that I have some personal and professional experience with. According to the American Psychiatric Association's *Diagnostic and Statistical Manual of Mental Disorder*, a diagnosis of ADHD requires *five* or more symptoms of inattention (or hyperactivity[10]) that persist for at least six months and that negatively affect social, academic, or occupational activities. The symptoms include: making careless mistakes, lacking attention to detail, having difficulties sustaining

---

10. In case you are interested, the symptoms of hyperactivity include: fidgeting, hand- or foot-tapping, squirming in seat, difficulty staying seated, running, climbing, or feeling restless when staying still is appropriate, excessive talking, blurting out answers before a question has been completed, difficulty with turn-taking, and frequent interrupting.

attention, having trouble listening, failing to follow through on tasks and instructions, being disorganized, avoiding tasks that require sustained mental effort, losing things, being easily distracted, and being forgetful. If you just read that list and thought—"Holy crap, that's me!"—you are not alone. After one of my brightest students with the most "streaky" productivity record was diagnosed with ADHD in graduate school, I started to wonder whether my husband, Andrea, and I fell "within the normal range" or not.[11]

Fortunately, the ability to pay attention is something I also study, from a "how brains do it differently" perspective. And as you'll read in the "Focus" chapter, "paying" attention is costly for any brain. But some people are clearly better able to stay on task, and resist distractions, than others are.

But here's the challenge—if I were to try to use my laboratory tests to sort people into "within the normal range" and "outside the normal range" buckets, my focus would be entirely on how *typical* a particular type of behavior is. Much like teachers who grade on a curve use class averages to weight a particular score—usually setting the average to a C grade—scientists can use statistics to decide whether a particular way of thinking, feeling, or behaving is typical or atypical by estimating how likely it is to be observed in their population of interest. Unfortunately, the choice about how to map "unlikely" to "abnormal" is a bit arbitrary. By convention, many scientists draw a line in the sand at the point where 95 percent of the population would fall in the "normal range" and the 5 percent with the most extreme values would be considered "abnormal."

But once that line is drawn, on either side of the cutoff you will

---

11. You might decide for yourself after considering the number of footnotes this book has—but on a more serious note, if you'd like more information, I highly recommend *Driven to Distraction: Recognizing and Coping with Attention Deficit Disorder* by Edward Hallowell and John Ratey—two experts in the field who both have ADHD and the clinical backgrounds to talk about it.

find two people who end up in *different* buckets even though their performance looks more similar to each other's than it does to that of most other people in their own bucket. One of them will wind up in the "within the normal range" bucket and the other will be "outside" of it. If you're the one who gets put in the "outside" bucket, you are more likely to get help, including access to services and treatments based largely on how most other people in your bucket work. But the person standing next to you, who got put in the "normal" bucket, may have very similar challenges without either the awareness or the resources to help. On the other hand, they don't inherit whatever "abnormal" label is associated with the bucket, and the stigma that can come with it.

Still, if I *were* to draw that arbitrary line in the sand based on the typical ways people perform on the attention tests in my lab, how well would they map onto the kinds of "real world" disturbances that the diagnostic manual considers? The short answer is not very well, and here's why: A person's ability to "resist distractions" doesn't exist in a vacuum. It lives in a brain with a whole host of other design features that may exacerbate, or compensate for, that particular one. And that brain exists in an environment with a particular set of demand characteristics that it may or may not be well matched for.

This explains why the diagnostic criteria for ADHD are more centered on *functionality* than they are on typicality. Rather than measuring how distractable a person is in the laboratory, the clinicians ask questions about whether a person's way of being "negatively impacts functioning." In fact, according to the CDC, about 9.4 percent of children in America are diagnosed with ADHD, and the numbers are steadily rising. If nearly 1 in 10 children has ADHD, it's not really that *abnormal*, is it? My point is simply that when it comes to how our brains are engineered, it's important to understand that *typicality*, or how frequently some design feature occurs in brains, and *functionality*, or how well that design feature is working for a person given the environment, are two different criteria that can be used to define "normal."

WEIRD science

To further complicate things, allow me to plant a seed (of doubt) about the role that culture has played in the historical definitions of both *typicality* and *functionality*. First, when it comes to *typicality*, both scientists and consumers of science alike need to ask themselves an important question: Do the people we study look like the people in the world that we are trying to make inferences about?

The answer to this question is almost always "no." As Joseph Henrich, a professor of evolutionary biology, and his colleagues so cleverly pointed out, the people we study—those on which the very definition of *typicality* is based—are WEIRD. That is, the *vast majority* of what we know about how people work comes from research conducted on participants from Western, Educated, Industrialized, Rich, and Democratic countries. Most of them are White college undergraduates. And if you spend as much time with undergraduates as I do, this might make you a little bit nervous.[12]

I'm not going to sugarcoat this. Much of the science in this book, including some of my own work, is based on WEIRD samples. This is clearly a limitation on how well I can help you understand how *you* work, especially if you're not WEIRD. We're doing our best to capture true neurodiversity in our lab, and if you're interested in lending your brain to science, or just learning more about it, please visit the research link on my website, chantelprat.com. Despite the glaring

---

12. Don't get me wrong, I like and respect the vast majority of the students I get to work with. But there are so many ways their brains are fine-tuned to their young-adult college bubbles that I have a really hard time accepting that they should be the prototypes for how all people work!

holes in the current research, I am confident that the foundational principles we discuss in this book—the biological spaces that brains can occupy, and the complex ways our environments shape and interact with these spaces—apply to brains from *all* walks of life.

But this brings me to my second point about the role of culture in defining the *functionality* of a given way of thinking, feeling, or behaving. The story of the bus drivers and the taxi drivers provides a straightforward example of the fact that the *functionality* of any brain's engineering depends on the context it operates in. I bet you can imagine a job in which "distractibility" might be quite functional—perhaps one in which you need to be able to detect unexpected changes in your environment and *adapt* accordingly? As you'll read more about in the "Adapt" chapter, these are the conditions our human brain likely evolved under—not the stable nine-to-five office or classroom life.

This is all just a roundabout way of explaining why this book is *not* going to tell you whether your brain is normal or abnormal, or even whether you're functional or dysfunctional. Even if I *were* interested in doing such a thing, I'm not qualified for that job. For the most part, the people I study in the lab operate in the "typical" bucket.[13] And while I'd like to think that the work I do in this space has implications for figuring out what it *means* when someone is defined as "abnormal," I also have to confess that I'd be totally OK living in a world without those buckets.

What if, instead, we tried to understand people as the multidimensional beings that they really are? While this kind of worldview would definitely make education, diagnosis, and treatment more difficult, there's little doubt in my mind that it would also make them more effective. As I hope the ADHD example illustrates, we all fall

---

13. I've also done a bit of collaborative work on autism spectrum disorder (note, there's a lot of variability in that bucket as well).

somewhere on many different axes of being. Sometimes, we might have extreme values in one dimension, but the extent to which that value is *problematic* depends on a lot of other factors, including our environments. And the reverse is also true—sometimes we have ways of thinking, feeling, or behaving that *are* problematic, but they don't come from one place. Instead, they can arise from multiple features that might be "within the normal range" in isolation but create a perfect storm when combined.

In this book, I'll define some of these axes in the brain, with the hope of helping you to appreciate the place you occupy in the multi-dimensional differences space. After all, Mr. Fred Rogers, who played a critical role in shaping *my* young brain, once said, "As human beings, our job in life is to help people realize how rare and valuable each one of us really is, that each of us has something that no one else has—or ever will have." So when the same brain heard Steven Pinker say that "all *normal* people have the same physical organs, and . . . we all surely have the same mental organs," it thought, *What a bunch of bullshit!*[14]

Besides, as Pharrell Williams says, "The same is lame."

What difference does it make?

To be fair, I don't think Pinker was trying to convince his readers that we are all *exactly the same*. Instead, I think his point was more about whether our differences are *relevant* or not, especially when they're viewed in light of our commonalities. "Differences among people, for all their endless fascination to us as we live our lives," he says, "are of minor interest when we ask how the mind works." If I set aside, for a

---

14. Though my higher, more objective self understands that we are both wrong in our own unique ways.

second, the fact that my entire career is based on this area of "minor interest," I can see his point.

To ground our two perspectives in the context of neuroscientific research,[15] allow me to introduce you to another nervous system, one that belongs to a nematode[16] called the *Caenorhabditis elegans*—or *C. elegans* for short. The entire nervous system of the *C. elegans* consists of a whopping 302 nerve cells, or neurons. These neurons, in turn, come into contact with 132 muscles and 26 organs. To state the obvious, the *C. elegans* is *not that complicated*. And while I think that even Pinker could get behind the idea that the difference between the wiring of *C. elegans* and that of our own brains *is* of interest when it comes to how minds work, a huge amount of what we know about how our own brains are engineered comes from studying simpler models. In other words, the differences between humans and roundworms are of little interest when it comes to how the brain works—at least at some level.

Say what?

At the end of the day, both nervous systems are information-detecting engines, built to collect data from the body and the environment and use it to make the best decision possible about what to do next.[17] To do so, they use many of the same mechanisms. Their basic processing unit, the neuron, is a marvelous cell with a clever way of accumulating evidence about what's going on in the world around it. When it does, it sends its own "summary" of the state of things on down the chain of communication. On the receiving end of each neuron are a set of branches, or "dendrites"[18] that reach to-

---

15. This is why I study brains, after all—they allow me to move from the philosophical realm to the more concrete reality I take comfort in.

16. Nematode is technically just a more fun way to say roundworm.

17. Yes, roundworms *do* make decisions.

18. The branching of a human neuron is much more intricate than that of *C. elegans*. Each human neuron may receive inputs from ten thousand other neurons,

ward the other cells in the vicinity and try to eavesdrop on their versions of what the state of the world is. The neuron accumulates evidence on a moment-to-moment basis, based on the number and type of signals it receives, until it reaches a threshold. And when *that* happens—boom! It joins the gossip circle, releasing its own chemical signals into the spaces where other neurons are eavesdropping on it. If you'd like to geek out on the specific process by which chemical signals open and close physical channels, which then change the electrical voltage inside the neuron and cause more channels to open, a quick search of *action potential* on YouTube will get you lots of cool animations. For now, suffice to say that the way it works in *C. elegans* and a human is basically the same.

HOW NEURONS WORK

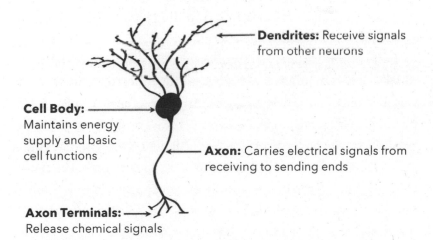

Dendrites: Receive signals from other neurons

Cell Body:
Maintains energy supply and basic cell functions

Axon: Carries electrical signals from receiving to sending ends

Axon Terminals:
Release chemical signals

In fact, there are enough shared features between the physiology of human and nematode neurons that hundreds of millions of

which would only be possible in *C. elegans* if it could connect to all the neurons in thirty-three of its closest worm friends.

national funding dollars are devoted to research on *C. elegans*. The things we have learned from this work fill dozens of books, with titles like *Neurobiology of the Caenorhabditis Elegans Genome, Ageing: Lessons from C. Elegans,* and my favorite, *WormBook.* Of course, if you consider the differences *between* human brains against the backdrop of how much we have in common with a *roundworm*, it seems easy to view them as insignificant.

But now consider the other end of the spectrum—the *differences* between the mental lives of humans and chimpanzees, our closest living relatives. As you might imagine, our brains are remarkably similar to chimp brains. This makes pretty good sense, if you consider the fact that the DNA blueprints that build human and chimp brains overlap by about 95 percent. But the *functional implications* of that 5 percent difference allow me to write a book, in a shared symbolic language that *you* can understand, while the wild chimpanzees still spend a good chunk of their day finding food and grooming one another to maintain their social bonds.

In this comparison, you can start to see that when it comes to the relation between minds and brains, a little difference can go a long way. But since you've never been a chimpanzee, here are a few examples that are closer to home. Remember the way you thought, felt, and behaved when you were a teenager?[19] Though the brain that guides you through life now still bears the scars of that time, the neural changes that happen across your life span can also have big consequences on your mental life. For an even subtler difference, consider how you feel first thing in the morning versus late at night. Within a twenty-four-hour cycle, changes in the neurochemical signaling of your brain's pacemaker, the suprachiasmatic nucleus, can have pretty dramatic effects on your inner workings. Hopefully, reflecting on the

---

19. To be clear, I am not trying to compare the brain of a chimpanzee to that of a teenager. And if you're still a teenager while reading this book, I hope it fuels your dynamic, growing brain with great information about how you work!

range of spaces your own brain and mind can occupy will help you start to appreciate the *relevance* of small differences. But before you decide whether these differences are *important*, let me talk a bit about their scientific implications.

Take, as an example, my early research investigating how the two halves, or hemispheres, of the brain collaborate to help you understand the stories you read or listen to. To get a better handle on the work your brain does for you in these situations, consider the following sentence:

The haystack was important because the cloth ripped.

Though this is a perfectly legal English sentence, you probably feel a bit disoriented after reading it. It's not that you don't understand the sentence, *per se*. You likely know the meanings of all of the words. And you can use your linguistic knowledge to figure out how the meanings of the words relate to one another. For instance, based on the order these words occur in, you know that it was the haystack, not the cloth, that was important. You also know that this importance is somehow *causally* related to the action of the cloth ripping. But you still don't really understand what the heck is going on.

This is because when we read, or even listen, to language, we have different levels of *understanding* it. The first is the one we've been discussing, based purely on the linguistic information contained in a sentence. But the second involves interpreting this information in the broader context of what you know about the world, and what's going on around you at the time.

The reason the haystack sentence feels *weird* is that it has been plucked out of its context. If I were to tell you that the sentence was part of a story about *parachuting*, how would your understanding of it change? Hopefully, things would click into place, shifting your comprehension of the sentence from a place of disconnected ideas to a scenario you can imagine, like a little movie clip unfolding in your mind. If it did, your brain connected a bunch of dots between things you already knew about the real world, like how gravity and parachutes work, and what was written on the page. From there, the reason a haystack might be important becomes clearer.

What's interesting about these two ways of understanding is that research on people with brain damage seems to suggest that different parts of the brain are involved in computing them. Prior to my research, it was generally believed that the left side of the brain, which is *typically* involved in processing linguistic information,[20] was responsible for understanding the ideas printed on the page; while the right side, which is typically implicated in more visual or spatial thinking, constructed the scenario. But these ideas, like most of what we know about how brains work, were based on findings that were averaged across groups of participants.

But we also know, from pioneers in the field of reading research like my graduate adviser, Debra Long, that not *all* people understand what they read in the same way. And I wondered whether this variance had anything to do with the way the jobs were divided up between the two sides of their brains. To explore this possibility, I conducted a study looking at differences in what each hemisphere remembered about a story, in more than 200 readers with different skill levels.

Here's how the experiments worked in a nutshell: Participants were told to read, and try to remember, short, two-sentence vignettes that were presented in the center of their computer screens. After reading a few stories, they would see a series of words flashed either in the center of their screens or just to the left or the right of the place they were told to focus their eyes. Their task was simple—to indicate

---

20. You'll learn a lot more about the division of labor between the two hemispheres in the next chapter.

with a button press as quickly as possible whether the word presented had been used in one of the stories. For instance, if I gave you the word "important" after reading the haystack story, you would say yes because that word was in the sentence.

Based on our participants' patterns of responding, we were able to reverse-engineer something about the way each of their hemispheres processed the stories. For instance, sometimes we presented words like "parachute" that weren't actually in any of the stories but were thematically related to them. If participants were slow to reject those words, or mistakenly said they had seen them, we had pretty good evidence that they were *understanding* the broader scenario of the stories. And we measured the linguistic type of understanding by checking to see whether people were faster to recognize words like "important" when they were presented after linguistically related words, like "haystack," than when they followed words from different grammatical clauses in the sentence, like "cloth."

And we used one last trick to figure out how each hemisphere might be involved in these different ways of understanding. Because of the way information flows from our eyes to our brains, everything coming from the left side of any focal point goes into the right hemisphere first, and vice versa. Though both hemispheres in a healthy brain eventually share this information, differences in the speed and pattern of responding to words presented to the left or right sides of the screen provide critical clues about how each hemisphere processed the sentences.

Although all participants in our study were college students with no diagnosed reading disabilities (in other words, they were all in the *typical* bucket), differences in their reading skill corresponded to brains that were doing different things—particularly in the right hemisphere. As predicted by the patient data, we showed that the left hemisphere of all of our readers *did* seem to understand the linguistic aspects of texts (that is, their left hemispheres understood

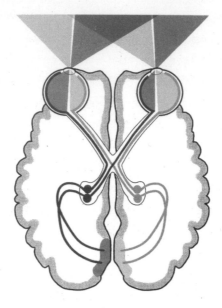

that the haystack was important, not the cloth). However, the right hemisphere of the less-skilled readers in our group was also sensitive to these linguistic relations. So much for language being a left-hemisphere-only function! And when it came to the scenario-based way of understanding stories, both hemispheres of the less-skilled readers got tripped up by words like "parachute," showing that they were sensitive to both the scenarios as well as linguistic features. On the other hand, only the left hemisphere of the skilled readers seemed sensitive to the scenarios. Ironically, the most-skilled readers had right hemispheres that were like Jon Snow from *Game of Thrones*— they knew nothing. They didn't respond differently when words like "important" were presented after "parachute" or "cloth," or "crow" for that matter. And they were no more tripped up by words like "parachute" that were thematically related to the stories than they were by words that were totally unrelated to the sentences.

At the end of the day, *no individual* in my experiment showed the specific pattern of results that was predicted based on the data you get when you average groups of people with different reading skill. It's kind of like taking a room full of people and saying that their average age is forty-two, even if no one in the room is actually forty-two. But in this case, the failure to understand how brains differed led not only to *incomplete* data but to *incorrect* conclusions about how the two hemispheres contribute to reading comprehension.

If you're still wondering why to care about this, imagine what might happen if you found yourself with an injury in the right hemisphere of *your* brain. What changes might the doctor tell you to expect? How might they evaluate the risks and benefits of some optional surgery?

Throughout my career, I have argued that although focusing on group averages has allowed the field to learn more quickly about the things we do have in common (like many of the mechanisms that underlie our sensory processes), it has also slowed our ability to understand the things that make us unique (like how we go about understanding stories, or jokes, or one another for that matter). One implication of this "one-size-fits-all" approach is that the vast majority of what we *know* about how the human brain gives rise to the human mind either ignores or glosses over the things that make us different.[21] For instance, many neuroscientists and even physicians still consider language comprehension to be the left hemisphere's job. And as a result, when it comes to understanding how, and in whom, the right hemisphere contributes to different ways of understanding language, there is little consensus in the field, even though descriptions of language difficulties following right-hemisphere damage have been reported for more than 150 years.

---

21. This is certainly not only true for human brains. Once, during a job interview, I asked a professor whether there were any individual differences in the brains of the genetically identical mouse population he studied. "Of course there are!" he said, a bit on the defense, "but we pretend they don't exist because it gets too complicated!" I didn't get that job.

But before I ride off into the sunset on my "differences matter" high horse, please allow me to make a confession: There are good, practical, reasons that people who are interested in human neuroscience don't study individual differences. The first relates to that whole "brains trying to understand brains" conundrum. The human brain is so incredibly complicated that we will definitely *not* fully understand it in my lifetime,[22] even if we *do* gloss over all the things that make us different and focus on our commonalities. The truth is that we still don't even fully understand *C. elegans*! Even though we have a perfect map of each of their neurons and what they are connected to, we cannot predict with 100 percent accuracy what a *C. elegans* will *do* in any given situation. We can get close, but we don't understand everything.[23] Imagine how that scales up from a map of 302 neurons to one of 86 billion, and you will have the appropriate mindset for appreciating how much we still don't know about *your* brain.

Which brings me to the second reason that studying individual differences in human brains is challenging. Many of the interesting variables can't be ethically manipulated in the laboratory. Instead, when a person walks in for testing, they bring all of their brain's design features with them—those that they were born with, as well as those that were shaped by their experience. But as you'll learn in this book, these things are often related to one another. Trying to tease differences apart to figure out *why* someone is the way they are is very challenging in the best of circumstances. The task always takes us back to one of the oldest questions in psychology—how much of what makes you *you* is inherent in your DNA, and how much has been shaped by your experiences?

---

22. No matter what Elon Musk says . . .

23. And this might be because we haven't done enough research on the ways *C. elegans* can be different!

The misunderstood battle of nature versus nurture

So what came first—the linguistically ignorant right hemisphere, or skilled reading ability? By now, most people who study human behavior understand that our biology and our experiences are so intertwined that it hardly makes sense to "blame" one or the other when trying to figure out what makes you *you*. The answer is always a combination of both. For one thing, every single life experience changes your brain. Some of the changes are inconsequential and others are incremental. But on rare occasions, for better or for worse, a single event can change the way we work, *forever*.

This is an important note to take, before diving deeper into the neuroscience of you. The fact that something about your brain *causes* you to think, feel, or behave in a certain way does not necessarily mean either that you were born that way or that it can't change. The truth is that your brain is a moving target. And most research linking brains to behaviors, like my work on the two hemispheres and reading skill, only looks at a single point in time—a snapshot, so to speak. With this kind of experiment, it's simply *impossible* to tell how much of a particular brain design is stable or has been shaped by your experiences.

One way to separate the influences of our genetic blueprints (or "nature") from our environments (or "nurture") is to do a longitudinal study. With this type of design, researchers measure the same brain at different points in time, to see how general maturation, or a specific experience, might change it. For example, in yet another clever follow-up experiment conducted with London's taxi drivers, Katherine Woollett and Eleanor Maguire did exactly this. Their goal was to figure out whether people who are able to pass the Knowledge test were born with bigger hippocampal tails, or whether it was the act of studying for the test itself that caused this area to grow.

To do this, they took images of the brains of 110 people on two occasions, three to four years apart. The majority of them (79) were taxi-driving hopefuls, first scanned while they began training to become taxi drivers but had not yet passed their tests, and the rest (31) were "control" participants, selected to match them on factors like age and IQ that might relate to the sizes and shapes of their brains. And since more than half of the trainees fail to pass the Knowledge, the researchers planned to make two comparisons with their data. First, they wanted to compare the brains of people who eventually passed the test to those who didn't, to see if there were observable brain design features that separated the groups. Second, they wanted to see if there were any notable changes that happened as a result of studying for the Knowledge and stuffing one's brain full of maps.

The results from Woollett and Maguire's longitudinal study provided pretty clear evidence about the causal relation between the taxi drivers' brains and what they'd been asked to do. Before training, there was no way to identify who would or wouldn't eventually pass the Knowledge. There were no reliable brain differences between the eventual "pass" and "fail" groups when they signed up for training—not in the size of their hippocampi, or of any other brain region for that matter. In fact, the only difference between those who passed the test and those who didn't was the amount of time they spent training every week. The group who passed spent an average of 34.5 hours each week training, while those who didn't typically spent less than 17 hours a week studying! And over three years, that intense training schedule left its mark, but only on the brains of the group that passed. After squeezing all of that knowledge into their brains, their hippocampal tails *grew*.[24] In other words, the excep-

---

24. In case you're wondering, this study did not show significant shrinkage of the head of the hippocampus. It's possible that the cost of this studying takes a bit longer to show up, as the same group has shown that the years of driving experience also influence the brain changes associated with driving a cab in London.

tional brains of the London taxi drivers were created by the demands placed on them. Case closed.

For researchers who don't have the time, money, or desire to follow their participants through life and take repeated measurements of their brains, twin studies provide another option for disentangling the influences of nature and nurture. The field of behavioral genetics has largely progressed this way—attempting to disentangle nature from nurture by looking at people who share different proportions of each. For instance, monozygotic, or "identical," twins are created from the same egg and sperm and are *nearly* genetically identical at birth;[25] whereas dizygotic, or fraternal, twins come from two different sperm and two different eggs, and thus share the same amount of genetic overlap as any two nonidentical siblings. Many studies estimate *heritability*, or the extent to which some measurable characteristic is genetically driven, by comparing how similar it is in pairs of monozygotic twins to pairs of dizygotic twins. If the monozygotic twins are more similar to each other with respect to the characteristic of interest (say, their ability to remember the location of landmarks) than dizygotic twins are, the difference between twins is assumed to be related to genetics. This type of analysis relies on the *assumption* that monozygotic twins and dizygotic twins share roughly the same degree of overlapping environments.

The problem with this assumption is that some characteristics that have genetic influences, like extraversion (which you'll read about in the "Mixology" chapter), also influence the kinds of environments and experiences people will seek out. And other genetic factors, like how tall or attractive you are, can influence your experiences by shaping the way others treat you. To further muddy the nature-versus-nurture waters, the rapidly evolving field of epigenetics is showing that

---

25. We'll talk a bit about epigenetics, and why they may not be perfectly identical, in a bit.

environmental experiences can create chemical changes in our DNA! As a result, a single gene can have different effects on the proteins it creates in the brain (or body) when placed in different environments. Through these mechanisms our experiences can become "biologically encoded."[26] In other words, if you put the same strand of DNA into different environments, it might build different people.

But sometimes the outcomes *aren't* that different.

The documentary *Three Identical Strangers* does a fantastic job capturing this. The movie is based on the remarkable true story of identical *triplets* who were adopted into different families at birth, and only discovered one another accidentally at the age of nineteen. In case you haven't seen it, I won't ruin the surprise (and sometimes scandalous) twists—but suffice to say that the ways in which these boys were similar to one another might go beyond what your mind conjures up when you think of how your biology makes you *you*. Of course they look, walk, and talk alike—but smoke the same brand of cigarettes? That's just wild. Or is it?

The only problem with anecdotal evidence like this is that we get so captivated by the story that we don't think about the facts objectively. For one thing, the similarities stand out, but the differences are easy to ignore. It's not that anyone would have been shocked if the triplets liked different kinds of beer[27]—but the fact that they all smoked Marlboros caught our attention. Which brings me to my second point about statistics and coincidences: To figure out how surprising the similarities between long-lost twins (or triplets) might be, we've got to ask the question, "How likely is it that any two random strangers who meet on the street would also be similar in this way?" When it comes to what kind of beer you drink, or what brand of cig-

---

26. Special thanks to Noah Snyder-Mackler for helping me wrestle with the concepts that underly the biological pretzel that is epigenetics.
27. And they may have, for all I know . . .

arettes you smoke, the answer depends on how popular that choice is. According to a market research article I found, in 1980 when the triplets met, Marlboros were the most popular cigarette with people in their age group, capturing about 40 percent of the market. It's still remarkable, but somewhat less remarkable than if they all smoked Camel Lights. To be scientific about the question of whether a person's taste in cigarettes can be genetically influenced, we'd need to look at a bunch of monozygotic twins, separated at birth, and see if the likelihood that they smoke the same brand is significantly greater than the likelihood that any two unrelated people plucked off the street would.[28]

I know, I'm no fun.

But the good news, when it comes to our discussion about nature versus nurture, is that I already had this kind of scientifically skeptical attitude on board when I met *One Really Freaking Similar Stranger* named Maia on April 7, 2020. There I was, in the middle of writing a book about how your brain makes you *you*, when I received an email from a twenty-year-old stranger with the memorable subject line "49.5% Match! (You might want to sit down)."

The first thing I noticed while reading the email was how much she "sounded" like me. Though her words were more carefully chosen than mine typically are, they were also a bit silly, and emphatic in a way that was *very familiar*. Unless something like this happens to you, you might never think you'd recognize yourself in the way someone uses an exclamation point. But I did![29]

---

28. Of course, this would be embedded in the statistic of whether they smoked in the first place. According to a twin study conducted by Jacqueline Vink, this depends on two factors: (1) the likelihood that someone will try smoking in the first place, which is estimated to be 44 percent genetic and 56 percent environmental, and (2) the likelihood that they will develop dependence on nicotine, which was estimated to be 75 percent genetic and 25 percent environmental.

29. To be fair, I have no idea how to go about finding statistics about how and when people use exclamation points.

The next things that struck me were the similarities in the things she chose to share about herself. Not knowing how I'd feel about being contacted, she strategically kept the email short and sweet. I imagine she thought quite a lot about what she wanted me to know about her, in case she never got the chance to talk to me again. Under those conditions, she chose eight things to share: (1) her love of singing, and the fact that she was studying to be a music teacher; (2) her love of animals, especially horses; (3–6) were brief mentions of hobbies, which included hiking, painting, traveling, and playing Mario Kart; (7) that she was voted "class clown"; and (8) that her Taco Bell order was a Crunchwrap Supreme with spicy potatoes and guacamole.

At this point, the feeling that I was talking to the twenty-year-old version of myself was through the roof. As you will probably figure out by the end of this book, I am also a *huge* animal lover. Hopefully, you're holding me accountable here and thinking, *Wait. What's the chance that two random strangers who meet on the street will also love animals?* And *that* would be a valid point. But I think I'm an outlier in how much I love animals. Like, I still go to petting zoos, even though my child is twenty-six—and I stay too long. When I was little, I brought a baby duck home from the feed store because it was cute. I named him Quackers and filled a wheelbarrow up with water so he could swim in my backyard.[30] In my adult life, I have become infamous for finding lost or injured animals, including Hugo, the dehydrated little baby raccoon I found in a gutter and raised in my garage until he was strong enough to be released. In my lifetime, I have had at least twenty different *types* of pets, starting with sea monkeys and an ant farm, and working my way up through fish and lizards in college, to finally fulfilling my childhood dreams and buying an off-the-track racehorse for myself for my thirtieth birthday.

---

30. Don't worry, I found him a suitable home on a larger property with a pond when he outgrew my backyard.

So what are the odds? According to the most relevant statistics I could find, 4.6 million Americans ride horses for hobby or sport. That's about a 1 in 71 chance of finding someone on the street who rides horses. But maybe that's not a fair estimate, since it's more popular with certain demographics than others.[31]

But what about the other seven things? Love of music? I'm an amateur drummer, but my daughter, Jasmine, did musical theatre throughout high school. Hiking? Definitely. Painting? I don't have the patience, but my mom, aunt, grandmother, and great-grandmother are all stellar visual artists. Traveling? Definitely, but that's pretty common for those who are capable. And Mario Kart? I've only played a few times, but I always lose—possibly because I like to pick the bathtub as my vehicle of choice. I was *not* voted class clown, but as you might guess from my Mario Kart vehicle, I'm also not a particularly serious person. In fact, my husband and I, who share the same preteen sense of humor, describe ourselves as "astronauts of stupid."

What's kind of funny, in retrospect, is that the most salient thing about the list of Maia's "fun facts" was her Taco Bell order. No, I am not about to tell you that I eat Crunchwrap Supremes with spicy potatoes and guacamole.[32] *That* would be wild. But anyone who spent time with me when I was Maia's age knows that Taco Bell was a huge part of my culture. To be clear, it's not the fact that we both liked Taco Bell[33] that blew my mind. It's that I would also probably have chosen to include my Taco Bell order in the "these are the things that you need to know to understand me" package. Suffice to say that

---

31. We could get into the weeds with statistics pretty quickly, but I soon learned that she rode the same discipline that I did when I started, and even rode the same breed of horse that I owned. However, within the discipline, the thoroughbred is quite popular.

32. Although Andrea and I have had quite a few of them since Maia introduced us to the idea. I'm not going to lie—they're delicious.

33. Depending on whether you count sandwich places or not, Taco Bell is the fourth or fifth most popular fast-food chain in the United States.

reading the email from Maia, and then watching the slideshow her parents had prepared for me, was an unforgettable experience. Though I knew she existed, it was an entirely different thing to watch the life of someone that was created from my DNA unfold across my computer screen.

Her origin story starts the summer before I began graduate school, when I decided to become an ovum donor.[34] It's a choice I'm proud of—one that allowed me to help an incredibly kind family that had trouble conceiving on their own while earning a bit of money to help support my own child, who was four years old at the time.

And here's where my nature-versus-nurture story takes an interesting twist. When it comes to shared experiences, my daughter, Jasmine—the best friend I gave birth to—and I are as close as it gets. We grew up together. Because I was only nineteen years old when I gave birth to her and was a single mom until I met Andrea twelve years later, Jasmine and I did *everything* together. When she was little, there were *months* at a time when we were *never* physically separated. As we went about the maturation process (she usually a few steps ahead of me), many commented on our similarity to the *Gilmore Girls*.[35] I can see it, except for the fact that I'm *way* less cool than Lorelai, and she's a *little* less nerdy than Rory. Oh, and we're real.

Like the Gilmore Girls, Jasmine and I overlap a lot in our "likes" (trash television, Zumba, Irish food, and '90s hip-hop, to name a few) and "dislikes" (anything even remotely scary, people who drive too slow, artsy films,[36] and having our feet tickled, for starters), but we

---

34. For those who don't know, this is the female version of a sperm donor—except, instead of getting a magazine and romantic time alone in a room with yourself, you get shot up with hormones for a month, and then your eggs get sucked straight out of your ovaries with a giant needle. It wasn't fun, but it was worth it.

35. *Gilmore Girls* is one of the best family-friends TV shows of all time. If you haven't seen it, I highly recommend it.

36. Though we are both growing in this area, with Andrea's influence.

have *very* different temperaments. She is chill (except when driving), and I am not. She's a deep, careful thinker, and I'm fast, spontaneous, and impulsive. While raising her, I never thought, *Jasmine is exactly like me*. I always thought, *We make a perfect team*.

Maia, on the other hand, seems to have a freakishly similar temperament. If the number of exclamation points in her email weren't a dead giveaway, most pictures of her hold some clue to our shared personality traits. We are both clearly high on the extraverted scale—I like to call it "pizzaz," but the kids these days might also call it "extra." Suffice to say that neither of us blends in very well. The other day, Maia sent me a picture of herself cruising around with Pepper, her pet bearded dragon, in this *giant* pink aquarium-backpack thing she bought so that he could come have adventures with her. Wow.

What might the similarities and differences I share with these two amazing young women with whom I also share half of my genes reflect about the role that genes and environments play in shaping our brains? In the pages that follow, I'll describe some of the ways that our brain design is influenced by nature and nurture independently, and how those two things interact. In Part 1, I'll focus on biological features. However, as you'll learn, even the smallest aspects of our biology are also shaped by our environments. When applicable, I'll talk about the heritability of different traits, or the percentage of variability that is *estimated* to come from genetic influences, based on twin studies and the like. Then, as we shift into Part 2, our focus will turn to the jobs we ask our brains to do, and how our life experiences and biology interact to shape the way we go about accomplishing them. Throughout this process, I have no doubt that you'll be thinking about how you came to occupy the "difference space" you currently inhabit, and I'll do my best to provide clues along the way. But before we go there, I'd like to add a few more words about what you should and shouldn't expect to find in the pages that follow.

You probably think this book is about you, don't you?

It's about time to address the elephant in the room—the fact that I haven't told you anything specific about how *your* brain works yet. But you're still here, which, I hope, is a sign that I've at least got you thinking about it. In the pages that follow, I plan to provide you with a solid foundation in the neuroscience of you—one that describes both how biological engineering differs across brains (in Part 1) and how the jobs they do provide the testing grounds for bringing out the differences between us (Part 2). Of course, to fit what I've learned in twenty-plus years into a book that wouldn't warp your brain like the Knowledge test, I had to make some tough choices about what to put in and what to leave out.

My decisions about what to include were largely driven by the aspects of brain design that can most easily be reverse-engineered. As a result, many of our discussions will be centered on characteristics like handedness or personality traits—things you either already know about yourself or that can be measured using the assessments you'll find in the book. But keep in mind, if you find yourself wanting to know more about how your brain works, feel free to check out the "Research" link on my website, chantelprat.com, at any time. There you'll find a variety of links to brain games that you can play to get more time-sensitive measures of some of your own brain's design features.

Whenever possible, I also chose to discuss topics that have been thoroughly studied, with multiple lines of converging evidence. Unfortunately, this is the exception rather than the rule with individual differences in research in neuroscience. Many of the experiments I describe have been conducted within the past five years. Try to keep this in mind as you read. This is a new field, and the cutting edge can also be the bleeding edge. I imagine that in another five years, what we know will have changed substantially. At least, I hope that's the case, because there is still *so much* about you that we don't know! Given where the field is at, my goal is not to give you all the answers but to give you the tools to be able to think about what we do and don't understand about how different brains work.

When it comes to what I'm *not* going to talk about in this book, one of the biggest topics is what makes one brain better, or worse, than another. It just doesn't make sense to me, even though I was born before the "every brain gets a trophy" generation. As the taxi-driver experiment illustrates, you've really got to be thinking about the match between a brain and an environment to decide whether it's a good fit, rather than talking about the absolute "goodness" of a par-ticular design feature.

For related reasons, I'm not going to spend a lot of time telling you how to *change* your brain. While I'm all for a growth mindset, I also think many of us would be better off if we could stop to understand and (dare I say it . . .) embrace the way our brains work. There's a rea-son they do the things they do, even if they drive us nuts (literally and metaphorically) in the process. Of course, I will talk to you about the kinds of experiences that may have gotten you where you are, and will occasionally provide little life hacks for things I think we could all use help with—like counteracting the effects of chronic stress on the brain. Still, at the end of the day, my hope is that your idea of what might be "better" or even "normal" will expand to incorporate more dimensions in the space of how we're different.

Another thing I'm *not* going to talk about is group differences, like the difference between the male and the female brain. Doing so is re-ally just a way of moving from the "one-size-fits-all" approach to the "one-size-fits-everyone-in-this-bucket" approach. It's not necessarily better. In fact, it can be a lot worse if not done thoughtfully, because things like "maleness" and "femaleness" are very strongly entangled in nature/nurture interactions. For example, from the moment a baby is born, adults use language differently with males and females. A baby's biology, from moment zero, shapes their experiences based on the expectations people have of them.

And even if you could separate nature from nurture when it comes to sex differences, the most frequently reported differences between male and female brains—things like the idea that females have more symmetrical brains than males do—are not consistently found in the literature. What this means is straightforward, if you ask me: For *any* brain design feature of interest, there will be differences between people, period. Deciding whether groups (say males versus females) are significantly different from one another involves using statistics to show that the differences within a bucket are smaller than the differences between buckets. This depends a *lot* on how many people are in the buckets, and how representative the people in the buckets are. As you're probably starting to guess, I'm not a big fan of putting people in buckets anyway, so we're just not going to go there.

And finally, a word about how I've chosen to report the science, and the scientists responsible for it, in this book. I hope I've already convinced you that brains are complicated, and by extension, that neuroscience is hard work. I believe that the people conducting this research are all trying their best to solve pieces of really difficult problems. I respect that act, in and of itself, tremendously. As a result, I've decided neither to use honorific titles nor to talk about the universities these scientists are working at. One practical reason for this is that it can be hard to tell if the person who wrote a research article has already gotten an advanced degree or if they're conducting amazing research while in training. I'd hate to get it wrong, but I'd also hate for you to think that if the first author of a paper doesn't have a "Dr." title, the paper isn't trustworthy.[37] This is also the reason I'm not

---

37. In case you think that you need to have a PhD to do high-impact research, part of my daughter Jasmine's master's thesis was published in *Science*, the most

going to tell you whether Author So-and-So is from an Ivy League school or not. Unless it's somehow relevant to the story, I don't think it should matter. Almost all of the research discussed in the pages that follow has gone through the peer-review process. This certainly doesn't mean that it's flawless—but it does mean that other scientists with relevant expertise have agreed that the science is sound. And most of this research is conducted by *teams* of scientists. As much as everyone on the team deserves credit, I think you would get really tired of reading paragraphs of names every time I described a study. So I made a choice to reserve the limelight for the first author on these studies, who—by convention—is the one who does most of the writing up of the research. Some authors prefer to refer to the work based on the most senior or most recognizable name in the group, but I wanted to be as transparent as possible when assigning credit.

Occasionally, I will mention details like how many participants were in a study. This *does* matter. All other things being equal, the more people in a study, the more likely it is that the findings from that study will withstand the test of time. And speaking of things being equal, while I would love to say that I'll report how *representative* the populations studied were, demographics other than age and sex are rarely reported. Unless there's something noticeably out of whack (like a study includes only males for no good reason), I probably won't talk much about the characteristics of the participants studied. But this is clearly a place where I'm hoping my field will grow.

So now that we've laid the foundation for being responsible consumers of neuroscience, let's roll up our sleeves and get to work learning about *your* brain. Because, as Brené Brown said, "People are hard to hate close up. Move in." And I can't help but wonder whether taking you *all the way in*, to the place where we're all pink and squishy,

---

prestigious scientific journal of all time. I've never gotten a paper published in that journal—but I'm still trying!

might help you to appreciate the nuances of yourself, as well as others who are different from you. Because in the hundreds of conversations I've had with friends, family, and strangers about my research, two things stand out: First, almost everyone is interested in neuroscience and the window it can provide to the self. Statements like "I'm not wired that way" show a lay understanding that something about the way your brain works makes you *you*. And second, many of us feel a little bit weird. After learning what I do for a living, you wouldn't believe how many strangers have told me, "You could write a whole book about my brain!" And as it turns out, they were right.

PART 1
BRAIN DESIGNS

How Differences in Brain Engineering Shape the Way You Think, Feel, and Behave

Bus rides are great exercise for the imagination. During the commute to and from work, my wandering mind often takes me places far away from my physical surroundings. Like the dreams I have at night, the contents of my daydreams vary from the fantastic (Jason Momoa is bringing me a drink with an umbrella in it. I can *feel* the warmth of the sun on my face.) through the mundane (Don't forget to send an email to so-and-so about such-and-such) to the horrific (Someone grabs the wheel of the bus and turns it sharply. We are careening toward the guardrail on the bridge that protects us from the water below.). In each scenario, the content of my conscious awareness, my *mental reality,* so to speak, has very little to do with the physical reality that my body is occupying.

Although I have been studying the neural basis of mind wandering for quite some time in the lab, I was slower to realize the broader, real-world implications of this ability to "pull the clutch" and let our minds run amok without making contact with our external reality. The moment its importance first clicked for me, I was on a bus ride like many others. On my way to work, I found myself mentally "rehearsing" what I feared would be an uncomfortable meeting with one of my students.

The student was falling behind, and I wanted to understand why, so that I could figure out how best to help. In my mind, I practiced different approaches to discussing the problem with them, hoping to find a "way in" that would come across as more caring than critical.

Around the third mental iteration of this "motivational talk," the expression of the woman sitting directly across from me on the bus caught my eye. Her soft-focus look clued me in to the fact that whatever she was *seeing* at the moment had little to do with our shared environment. My thoughts of the conversation ahead melted away, and my mind became captivated by the realization that although our bodies were in roughly the same place on Earth at the same time, our minds were engaged in very different journeys. I tried to imagine what she was thinking about, and felt comforted by the fact that my worries, which seemed utterly central seconds before, were totally invisible to her.

It was as if we were each riding the bus with giant bubbles around our heads. Inside these bubbles, private screenings of our personal "reality shows" were being projected. Of course, in my bubble I was the star of the story, playing the role of a well-intentioned scientist, known to veer, occasionally, into the overly critical lane. In hers, I was at most an extra, filling the seat on the bus across from the lead actor. With a quick glance around me, I realized that this shared moment was part of a different scene, in a different story, for every rider on the bus. Realizing how independent our mental experiences were filled me with the same sense of perspective I get when I gaze at a sky full of stars. And in that feeling of smallness, I was reminded of the vast gap between *my* reality and *the* reality.

If there is one thing I hope you'll take away when learning about the neuroscience of you, it's this: You are neither an actor in nor the passive observer of your reality. You are the *creator* of it. In fact, if one were to define your conscious awareness as the movie being projected inside your bubble, your brain would be the projector, the director, the production team, and the audience all in one! And though my aha moment was centered on the fantastic worlds that the wandering mind creates, the first section of this book describes the ways that different brains can create different story lines, even when they're focusing on the same "ground truth."

In this part of the book I will describe some of the different biological features that shape the way your brain creates and produces the stories you experience as your personal reality. First, in "Lopsided," we'll discuss how the two halves of every brain come up with slightly different stories about what's going on in the world, and how variability between people can be driven by the divide within. What might being left-handed tell you about how the two sides of your brain see things? This chapter will describe the truth behind the common myth about what it means to be a "left-" or "right-brained" thinker. Then we'll move to "Mixology" and talk about the roles that the ingredients of our neural cocktail play in our brain's communication system. If you want to know what being extraverted has in common with a cup of coffee or tea, this chapter may be of interest to you. And finally, in "In Sync," we'll cover the way your brain uses neural rhythms to coordinate the chorus of signals traveling around in your head at any given time. And as you'll learn, some of our choirs have more bass than others. In this last chapter on brain design, I'll describe the way differences in your brain's preferred neural rhythms influence the way your brain samples the world "out there" and creates its stories by connecting the dots.

Taken together, these chapters will provide critical insights on the way your brain creates the story of you. Because, as Brian Levine writes in his article on autobiographical memory and the self, "A good storyteller weaves setting, players, antecedents, story line, and implications together in a tapestry." And man, is your brain a good storyteller. The goal of this first section of the book is to give you some clues about how *your* brain's design features shape its storytelling processes.

CHAPTER 1

LOPSIDED

The Two Sides of Your Brain's Story

If I were to show you a picture of your brain, the first thing you'd probably notice is that it looks like a big walnut (no offense), with two largely independent halves, or hemispheres, connected by a high-speed core. As strange as this might sound, it isn't a very unique brain design. In fact, all vertebrate animals have brains that are divided down the middle, and they have probably been engineered this way for hundreds of millions of years.

What makes human brains remarkable in this design space is how *lopsided* we are, on average. Differences in the size, shape, and patterns of connectivity in our left and right hemispheres leave us far from symmetrical. And as you'll learn in this chapter, these structural differences shape the way each side processes the information it receives.

However, contrary to the popular notion of the "left-brained" analytical person and the "right-brained" creative type, the most striking distinction between human brains isn't which hemisphere is "in charge" of things. Instead, differences in our characteristic ways of thinking, feeling, and behaving are driven by our degree of lopsidedness, or how big the differences between our two hemispheres are.

And so, this book about the differences between brains will begin with a discussion of the fundamental divide *within* them. But before we get into the nitty-gritty details about how *your* brain looks, let's talk about why evolution might have landed on the different options in the first place. The idea, in essence, boils down to specialization.

The costs and benefits of brain specialization

To better understand the pros and cons of having a more lopsided or balanced brain design, let's imagine that your brain is a team made up of two people. If both members of your team are well rounded, and have comparable skill sets, it would be easiest, and most equitable, to distribute tasks between them randomly. On the other hand, if one member of your team has incredibly strong verbal skills, while the other is an excellent graphic designer, your team would perform better as a whole if the tasks were systematically assigned to the individual best qualified for the job.

Job assignment in the brain works a bit like this. If the two hemispheres were truly identical to each other, there would be no rhyme or reason to which functions they might come to perform. But as soon as they start to differ—even a little bit—an opportunity is created for one hemisphere to be better suited for certain types of jobs than for others. When this happens, the assignment of jobs across hemispheres becomes more systematic. And as the jobs that a particular brain region is asked to do become more similar to one another, that region can adapt, developing a more specialized structure that allows it to perform the particular type of tasks it's involved in even better.

I assume that the benefits of specialization are somewhat self-explanatory. If everything else were equal, many people would rather have an extremely talented graphic designer on their team than one with average skills. But what if that graphic designer was bad at everything else? If your whole team were made up of people with nonoverlapping skills, what would happen if someone needed help or called in sick? One of the measurable costs to specialization in the brain is that the refinement process by which an area becomes specialized makes it better and better suited for doing fewer and fewer things.

Stefan Knecht and colleagues demonstrated this increased vulnerability associated with lopsidedness in a study that looked at language *laterality*—a term neuroscientists use to describe the extent to which any of your brain's functions come to depend more on one hemisphere than the other. To do so, they first measured changes in blood flow in the two hemispheres[1] when 324 volunteers named pictures in the lab. Then they selected 20 participants who had different patterns of laterality for speech, with approximately equal numbers of people who relied on their left or right hemispheres uniquely, or on both, for speech production.

Next, to study their vulnerability to brain injury, the research team used a tool called transcranial magnetic stimulation, or TMS for short. TMS uses magnetic fields to safely, and *temporarily*, stimulate different regions of the brain noninvasively.[2] And if you stimulate one area over and over for a long enough time, it runs out of gas[3]—creating an effect called a "virtual lesion." If you've ever gotten a blind spot after seeing a bright light, you've experienced a similar phenomenon.

As expected, when Knecht and colleagues created virtual lesions

---

1. This is an indirect measure of how much each hemisphere is contributing to the job at hand—kind of like fuel consumption is an indicator of how hard a car's engine is working.

2. This is just a fancy medical term that means we don't poke any holes in them.

3. We'll talk about the mechanism behind this in the next chapter!

in the hemisphere that a person's speech was dependent on, their participants got significantly slower at the language task they were asked to perform. However, the more balanced a person's speaking profile was—that is, the more *both* of their hemispheres were involved in the act of speaking—the less their behavior was affected when only one side of the brain was fatigued using TMS. The effect is kind of like benching different members of your team and measuring the dip in productivity that results. The more balanced brain designs, like the well-rounded teams, were more resilient to the injury of any single player.

But even for the majority of us who are lucky enough to make it through life without damaging too many brain cells, there are still prices to pay for brain specialization. One of them relates to how our hemispheres become different in the first place. Though I spent a decent amount of time in "Introductions" explaining how evolution has worked to cram as much brainpower as possible into our heads, the mechanisms that cause our hemispheres to become specialized may be an exception to this rule. According to the Right-Shift theory proposed by Marian Annett, the human propensity to be lopsided may be driven by a genetic variation that *shrinks* parts of the right hemisphere. According to Annett, our brains evolved this type of handicapping system as a way of refining job assignment in the brain.[4] Consistent with her theory, Annett's results suggest that people who have more "balanced" brains might not be as skilled at the more newly evolved human functions—like language—but they are also using *more* of the real estate in the right halves of their skulls, which you'll learn is important for many other things, like visuospatial skills. On the other hand, she argues that highly lopsided people are less likely to have deficits in language-related skills but are more

---

4. The difference in size is not apparent in all areas of the brain, and it's also not the only difference between the hemispheres, but we'll get to that more in a bit.

likely to struggle with the types of jobs that typically get assigned to the right hemisphere, like visuospatial tasks.

And there's one other thing I'd like you to keep in mind when considering the costs and benefits of the specialization of our two hemispheres. As you'll learn in this chapter, one of the ways your brain becomes specialized is by using highly experienced processing centers called *modules*. These modules are singularly focused on the task they've been given, and don't consider input from other brain areas while they are doing their jobs. The result of this is that a more specialized brain tends to process the world by piecing together specific details rather than taking the whole picture into account. In other words, as a brain moves from being more balanced to more lopsided, its processing shifts from focusing on the more global, "forest-level" features to focusing on more specific, "tree-level" details. We'll talk more about the specifics of this in the second half of the chapter. First, let's get to work figuring out how lopsided you are.

Assessing laterality

One of the best ways to determine how lopsided your brain is, is to measure a bunch of different functions in each hemisphere separately. If your left and right hemispheres do them equally well, your brain is likely more balanced, but if one hemisphere tends to take the lead on these functions, your brain is probably more lopsided.

We'll start with one of the most obvious asymmetries to observe in most people—our hand preference. Those of you who work with your hands for a living, or have suffered an injury that prevents you from doing so easily, are likely already aware of how much skill goes into precision hand movements. The rest of you might be largely oblivious to one of the most important benefits that our genetic differences from chimps created—our long thumbs. The fact that we can press our thumbs to the tips of each finger with precision levels of force allows us to execute movements ranging from removing an eyelash from someone's cheek to hitting a nail on the head with a hammer. And these common tasks probably require *a lot* more brainpower than you think.

In fact, the neural circuitry that controls the movement of your hands is so large that it creates a U-shaped bulge in your brain called the *hand knob*.[5] With a bit of training, you'd be able to identify your hand knob when looking at a picture of your walnut-shaped brain. It sits near the top of your motor cortex, a strip of brain that runs from temple to temple (about where a pair of glasses would fall if you rested them atop your head), and controls the movement of all of your body parts. In most people, you can even figure out whether they're left- or right-handed by comparing the size of the two knobs in each hemisphere. And this is how we're going to start the process of reverse-engineering *your* brain.

Though most people identify as either right- or left-handed, handedness is not a binary category. Instead, we each fall on a continuum ranging from extremely right-handed to extremely left-handed. Figuring out where you fall along this axis is the first step to understanding how lopsided your brain is. To start, I'll give you a questionnaire I adapted based on the Edinburgh Handedness Inventory. This simple checklist, which asks about how you use your two hands for everyday tasks, is by far the most common tool used by neuroscientists to measure handedness.[6]

To get an idea of where you fall along the handedness axis, answer each of the ten items below based on everyday activities that you might engage in with either your left or right hand. For each action,

---

5. This bulge is an example of the gyrification process that squeezed as much computational real estate as possible into a medium-size head.

6. However, if you'd like to get a more refined idea of the relative skill of your two hands, check out the "hit-the-dot" game on the research tab on my website.

answer on a scale ranging from +2 to –2: If your preference for this activity **is so strongly right-handed** that you wouldn't ever use your left hand, answer **+2**; if you **prefer to use your right hand** for this activity, but may occasionally use the left as well, answer with a **+1**; if you are truly indifferent, and **use both hands equally well** and equally frequently to accomplish this task, answer with **0**; if you **prefer to use your left hand** for this activity, but may occasionally use the right as well, answer with a **–1**; and finally if your preference for this activity is **so strongly left-handed** that you wouldn't ever use your right hand, answer with **–2**. The only time you should leave a question blank is if you have no experience with the activity in question (and if you've never held a broom, or a *toothbrush*, I'll do my best not to judge you, since it's antithetical to my goals for writing this book).

HANDEDENESS ASSESSMENT

1. Writing with a pen or pencil.
2. Hammering.
3. Throwing (most commonly a ball but any object will suffice).
4. Holding the match when striking a match.
5. Holding a toothbrush when brushing your teeth.
6. Using scissors to cut.
7. Cutting with a knife (without a fork, such as when chopping food for cooking).
8. Eating with a spoon.
9. The upper hand when holding a broom to sweep. (If it's been a while, grab a broom—sweep for science!)
10. Opening the lid of a box.

Now, let's calculate your handedness index. To figure out your "average" response, add the answers to each of the ten questions together and divide their sum by ten. To check your math, the result should fall within the –2 (strongly and consistently left-handed) to +2 (strongly and consistently right-handed) range. The closer you are to the extreme ends of this distribution, the more lopsided your brain is. Those of you who scored closer to the middle (between –1 and +1), the mixed-handers, likely have more balance in the capabilities of your two hemispheres. Still, you probably identify as right- or left-handed based on your answers to the first few questions. As you move from the top to the bottom of the scale, the precision required to execute the movements generally decreases, opening up the possibility for a less-skilled hemisphere to do a "good enough" job.

So, what does your degree of handedness tell me about how lopsided your brain is? The first thing to note is that the motor cortex in the left hemisphere of your brain controls the right half of your body, and vice versa.[7] If you are strongly right-handed, it is likely that the motor cortex in your left hemisphere, particularly around the hand knob, is bigger. The reverse is true for the much smaller percentage of the population that is extremely left-handed. We'll talk about the broader implications of what this means for how you work in a bit. For now, let's check out some other functions, to see whether your brain is consistently more balanced or lopsided in its job assignments.

For starters, let's check in with your feet. Although our feet are

---

7. I often find myself doing something that looks like a nerdy version of the Macarena when trying to remember which part of my brain controls what. Take the right hand and touch it to the left side of the top of your head where the hand knob is, then repeat with the left hand to the right side of the head. These are the motor cortices controlling the opposite sides of your body. Then, if you unfold your two hands and stretch them out in front of you, you have a model of what the two hemispheres see! Because, as you learned in "Introductions," the left side of the brain sees the right side of the world first, and vice versa.

much less skilled than our hands, most lopsided people will also exhibit a preference for using one foot over the other when executing skilled movements. Which foot do you usually kick with? When going up stairs, which foot do you typically lead with? What if I asked you to put the tip of your toe on a quarter? Would you instinctively pick one foot over the other? Most people will find these foot skills more interchangeable than the hands, but if you answered each of these questions consistently with one foot, it provides further evidence that skills are unevenly distributed in your two hemispheres.

Now, let's switch to an even more subtle function—the difference between how you use your two eyes. Though both eyes carry information about the world to the brain, some of us rely more on information coming in from one eye than the other. And here's a fun fact—most people have a preference for information coming in from their right eye![8] We might go about assessing your eye dominance like we did handedness by asking questions like which eye would you use to look into a microscope or the viewfinder of a camera? But we can also measure this a bit more objectively with the following "sighting" experiment: Find an object about eight to twelve feet away from you and hold one of your index fingers up in front of it. With both eyes open, you might have the experience that you can "see through" your finger, or you might feel like you see two fingers (depending on where you're focusing), but do your best to focus on the object and position your finger so that it is in a straight line between you and the object. Now, close your left eye. What happened? If your finger now looks like it is solidly blocking the object, you are *right-eye dominant*. If your finger now looks like it is off to the side of the object, try closing your right eye. Is it lined up now? If so, you are *left-eye dominant*. As

---

8. The distribution of "eyedness" is much less right-biased in the population than handedness, however. About 2 out of 3 people prefer their right eyes, while 9 out of 10 prefer their right hands.

long as you've picked something sufficiently far away, if your finger doesn't line up when you close either eye, you've got *mixed-eye dominance.*

By now you should be starting to see a pattern. Those of you who are strongly lopsided are more likely to find yourselves consistently preferring to use one side of the body than the other. Others, with more balanced brains, are both more likely to have mixed preferences within any body part and more likely to bounce around with respect to the sides they prefer on different body parts. But now, let's try a totally different kind of measure—one that assesses how your two hemispheres might *understand* the world in similar or different ways.

Take a look at the two faces below. Which looks happier to you?

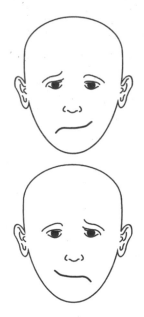

If you're thinking that this is a trick question—because they are the same face presented in mirror reversal—you are correct . . . but try it again with less thinking and more feeling. If you focus your eyes on the center of each face, does one seem happier than the other?

These chimeric faces are often used in research to figure out how the two hemispheres respond to facial expressions of emotion. They rely on the way the eyes are wired, which I mentioned in "Introductions." Information to the left of your nose goes to the right hemisphere first, and vice versa. So if you picked the face on the bottom, your brain is using more of the information processed by the right hemisphere to make its decision. But if you picked the face on the top, your brain is relying more on the left hemisphere to do these jobs. Of course, for those of you with the most balanced brains, the faces may have felt truly similar, and you may have just made a random guess about which was happier. Laboratory tests that use these kinds of faces to assess laterality typically show participants many different faces so they can tell how consistently a person depends on one hemisphere versus the other. For now, you'll have to rely on your intuition, imperfect though it may be.

Taken together, information from these assessments can provide a pretty good idea about how different the two hemispheres of your brain *look* from each other. In the sections that follow, we'll discuss some of the research that has examined how differences in lopsidedness relate to how your brain comes to understand the world around you. But before we do, let's talk a bit about how frequently the different patterns of results occur. Among other things, this information will allow you to understand how likely it is that any study you read about "average brains" will be representative of how *your* brain works.

How typical are you?

Though 90 percent of the population identifies as right-handed, only about 60 to 70 percent of people are strongly, consistently right-side dominant for motor tasks. Those of you who are will have scored strongly right-handed (close to +2) on the checklist, and will most likely use your right foot and right eye for skilled tasks. If you fall in this category, you also probably picked the face on the bottom as happier. The reason I can guess this is because you are in the majority group, which means that much of what we have learned about how jobs get assigned to the two hemispheres of the human brain will also be true of *your* brain. However, this isn't always the case. As you may remember from my research on reading in the two hemispheres, sometimes what we believe based on group averages doesn't accurately describe anyone.

And the waters get much murkier in the second-largest group of you, the 25 to 33 percent with more balanced brains. You might have driven yourselves crazy with these laterality tests, struggling to decide which hand or foot you might use for various tasks, and finding that your damned finger bounces all over the place when you close either eye. I apologize, but I think it'll be worth it to know this about yourself. After all, we've still got a lot to learn about your brain, particularly because neuroscientists (myself included) have done a crappy job deciding how to define you. Many of you will identify as right-handed. After all, the world is largely engineered for right-handers, so if your left hemisphere is capable of controlling your right hand, you've probably trained it pretty well. But some brain researchers consider anyone who isn't strongly right-hand dominant to be left-handed, while others consider anyone who isn't consistently left-hand dominant to be right-handed, pushing the mixed-dominance group in with the righties. Too often, researchers make some arbitrary cutoff in the middle or consider only the hand you write with to understand differences in brain laterality. To think that as many as 1 in 3 people are randomly classified in studies of handedness gives me the goose bumps!

Despite these inconsistencies, researchers like Stefan Knecht, who *have* been thoughtful about studying handedness along a continuum, tend to find that mixed-dominant people, particularly those who tend to use the right side of their body for most things, have brains that resemble the majority group on average, with a few more surprises thrown in. That is, if the right hemisphere is dominant for a task like face processing in the majority group, *your* right hemisphere is also probably more involved in face processing than your left is. As a result, most of you still probably picked the face on the bottom as being happier. But it's also likely that your left hemisphere is better at understanding faces than the left hemispheres of most strongly lopsided people. So the decision about which face to pick was probably harder for you. If I would've done it in the lab, I might have found that you were slower to make the choice. And if you have a balanced brain that slightly, but not strongly, prefers the left, this was probably even more true for you. In short, the more balanced your brain is, the more likely its job assignments will involve both hemispheres to some degree. We'll get into the details of what this really means about how you work in just a bit.

And this brings me to the rarest group, the 3 to 4 percent of you who identify as *consistently left-hand dominant*. Those of you who showed an extreme left-handed preference (close to –2) on the checklist may be just as lopsided as the strongly right-handed group. You most likely also prefer to use your left foot and left eye, and are more likely than either of the other groups to think the face on the top was happier! I don't want to play favorites, but this group is near and dear to my heart—and it's not *only* because I like to understand people who are different. It turns out that the single brain I have collected the most data from comes from an extreme lefty whom I have been able to test repeatedly over the past twenty-four years—my daughter, Jasmine.

In fact, my first job in neuroscience involved getting children to wear these annoying swim-cap-style hats with electrodes sewn into them that allow us to record electrical activity coming from their brains. And anyone who has ever tried to get a toddler to keep something on their head as part of a Halloween costume or something like that knows that this is probably one of the hardest gigs in neuroscience! My big advantage in getting the job was that I had more experience with toddlers than most of the undergraduates at UC San Diego. I had my own![9] And because she had a very easygoing temperament, I often brought Jasmine into the lab to practice my "capping" technique.[10]

The first time I looked at the brain data recorded by Jasmine's electrode cap, however, I was convinced that I had done something wrong. When she listened to words that she knew and words that she didn't, the differences in her brain activity (called the N400 because they are Negative deflections in electrical polarity that happen about 400 milliseconds after hearing a word) were larger over the right side of her head than the left. While some of the babies we were studying, primarily the younger ones or later talkers, were prone to showing the changes over both sides of the brain, none that I had seen showed this pattern of selective right-hemisphere sensitivity to words. To follow up, my supervisor, developmental cognitive neuroscientist Debbie Mills, suggested that we run some more tests, including an "oddball" paradigm in which the subject listens to a series of tones presented at one pitch, which are occasionally interrupted by a tone of a different pitch. In *most* people, this

9. I am eternally grateful to Debbie Mills, the principal investigator of this lab, for seeing the fact that I had a child as a plus and not a minus and giving me the chance to get my foot in the door.

10. In case you're judging me for using my child as a guinea pig, let me just point out that there are worse things a single mom could do to keep their kid entertained while working . . . and she grew up to be a scientist, so it couldn't have been all that bad!

creates a change in activity called a P300 (a Positive change in polarity that happens about 300 milliseconds after the tone is presented), which is bigger over the *right* hemisphere. In Jasmine, it was also flipped.

One of the coolest things about all of this is that Jasmine's brain told me she was left-handed before her body did! Although preferences *can* be detected earlier, most children start to show consistent hand preferences between eighteen months and two years of age. Jasmine was seventeen months old when she had her first brain-recording session, and once I realized her brain was flipped, I almost immediately noticed her strong preference for using her left hand. Over the years, I've continued to find this pattern in Jasmine's brain structure and function. The jobs of her hemispheres do not seem to be *randomly distributed*. Instead, her brain tends to show specialization that reflects the opposite patterns of what most of us consider to be "normal."

Unfortunately, extreme lefties are often *left out* of neuroscience research, as a consequence of the one-size-fits-all approach to brain science. The justification for this practice is that left-handers (loosely defined) are "more variable," and that if we average data from brains like Jasmine's with data coming from *typical* brains, it makes a mess. As a result, we don't know nearly enough about those of you who didn't grow up in neuroscience labs.[11] However, the few published studies that have looked at this systematically have found similar results to my own observations of Jasmine—though it is rare to have reverse lateralization of your brain's jobs, these patterns occur most often in *extreme* left-handers.

As my experience as a neuroscientist and a mother have grown in tandem, I've often wondered whether some of the weird things

---

11. I just wrote a grant proposal to try to do my part to fix this. Here's hoping I can convince the funding agencies that it's not acceptable *not* to understand how left-handers work.

about Jasmine—like the fact that she turns her head to the left and watches television out of the right corners of her eyes (which puts more of the information in her left hemisphere first), or the fact that she is both highly intelligent and not very *fast* in terms of information-processing speed—have something to do with her rare brain organization. In the section that follows, we'll talk about what we do know about why jobs get assigned to one hemisphere or the other, and what it might mean for those of you whose brains are different from the majority.

From structure to function: How jobs get assigned to the two hemispheres

To help you understand the complicated relationship between how your brain looks and how it works, I'd like to explain something that too many people in my field get wrong—the distinction between a brain *function* and a brain *computation*. If we return to the metaphor of assigning jobs to people on a team, you might think of a *function* as the job that gets assigned, and the *computation* as the skill set that allows a person to perform the job well. When it comes to understanding how the brain works, both scientists and laypeople too often describe what brain regions do in terms of their *function*, without understanding the more fundamental *computation* that allows an area to contribute to that particular function. But if you want to understand why your degree of handedness relates to which side of a face you pay more attention to, we're going to need to dig deeper and discuss the computations that link the way your brain is designed structurally to the functions it contributes to.

Let's use language—one of the brain's most important and impressive *functions*—to illustrate my point. Although I have spent much of my life studying the ways that *both* hemispheres contribute to language processes, most people consider it to be the poster child for a lateralized function—one that primarily gets assigned to the left hemisphere. In fact, it was French physician Paul Broca's description of a patient who seemed to lose only the ability to speak following damage to his left hemisphere that launched the whole notion that mental functions could be assigned to specific brain regions in the first place. More than a hundred and fifty years after his discovery, almost every textbook that covers language in the brain will point to a region in the left frontal lobe, now referred to as "Broca's area" and label it with the function of "speech," and another, slightly above and behind the left ear, with the function of "language comprehension."

But here's the deal. Your ability to *use* language—the system that allows you to translate ideas into and out of the arbitrary symbolic forms we use to communicate them—relies on many different types of computations. And factors like whether you're on the sending or receiving end of a language exchange, or whether you're using verbal or written symbolic systems, influence the types of computations your brain relies on to perform the function. The extent to which speaking and comprehending language actually recruit separate brain regions depends on which of their many underlying computations you're interested in.

Let's use the ability to produce speech, which was impaired in Broca's patient, as an example, since it kicked off the whole "mapping functions to brain areas" approach in the first place. When Broca's area is damaged, most traditionally lopsided people experience difficulties producing speech. But that doesn't necessarily mean that the *function* of Broca's area is speech. That would be like inferring that if a flat tire prevents you from cruising down the road at top speeds, the function of your tire is to propel your car forward. To produce a meaningful utterance with your mouth, your brain needs to execute a complex sequence of computations that first translate an idea in your head into the linguistic symbols used in your language for expressing them. It then needs to link those linguistic symbols to the programs that generate the elaborate sequence of movements—an oral ballet in which your tongue, lips, teeth, nose,[12] and vocal cords do just the right thing at the right time—to shape the air you exhale into a form that another brain can "understand" when the resulting vibrations hit its eardrums.[13]

And as those of you who have had experience driving "well-worn" cars probably understand, there are *a lot* of different conditions besides a flat tire that can prevent a car from moving forward. In fact, a more accurate model of how cars work will tell you that many different things need to go well for your car to cruise down the road uneventfully. Speech is also like this, and scientists like Nina Dronkers and her collaborators have shown that a *different* region of the brain, the insula, may be even more important for fluent speech than Broca's area.[14]

To further complicate the matter, impaired forward motion is not the only thing that happens when you get a flat tire. It's also more difficult to turn, and the ride gets a lot bumpier. Similarly, if you look carefully enough, you can observe a variety of linguistic and nonlinguistic deficits following damage to Broca's area. For example, some people lose the ability to use word order to *understand* the meaning of a sentence, and others have difficulty understanding what actions are depicted in images!

The point I'm trying to make here is that 160 years after Broca's

---

12. Of *course* you speak through your nose. If you don't believe me, try to plug your nose and say "nose." Sounds like "doze," right?

13. The fact that most brains do this so effortlessly is nothing short of a freaking miracle to my mind, given the complexity of information processing required.

14. In fact, Dronkers and her team were able to use modern brain imaging equipment to study the preserved brains of Broca's first patients. When they did, they found that the damage was more extensive than reported by Broca and included the insula. I therefore propose that we start calling the insula "Dronkers's area" to celebrate the woman who got it right 130 years later.

observation, much of what we understand about how brains work is still related to the *functions* that get disrupted in *most* people when their brain becomes damaged, or to the tasks that make it more active in experiments conducted on healthy brains. But to really understand how *your* brain works, we'll need to be able to discuss the mechanics. What is it about how the left and right hemispheres of your brain are engineered that allows one of them to be better suited for a particular function than the other?

Who am I speaking with? Two sides of language in the brain

One of the biggest clues we have about why jobs get assigned to one hemisphere or the other is the relationship between handedness and laterality for language. The fact that most people prefer to use their right hand, which is controlled by their left hemisphere, and *also* rely on the left hemisphere to produce speech, suggests that something about the way the left hemisphere has evolved makes it better suited for the computations that both functions rely on. Because Broca's area neighbors the part of the brain that is responsible for moving the lips, mouth, and tongue, many people assume that this shared computation relates to motor coordination, or how accurately your brain can puppeteer your body.

But not *everyone* uses their "preferred" hemisphere for hand control to speak. Stefan Knecht and colleagues demonstrated this when they recorded differences in the extent to which 326 people with a range of handedness relied on their left or right hemispheres to produce speech. In this study, which was the prequel to the TMS experiment described at the beginning of this chapter, participants were sorted into seven groups ranging from consistently right-handed to consistently left-handed based on the same handedness inventory you took. Because Knecht was interested in *understanding* lefties rather than excluding them, his sample had a larger number of consistent left-handers (57) and relatively balanced-handers (101) than would be expected if you sampled randomly from the population. And when he compared the blood flow in the two hemispheres associated with speaking in the seven groups, he found striking differences. In the consistently right-handed group, 96 percent of people had greater blood-flow changes in their left hemisphere than in their right when speaking. In other words, almost everyone who strongly prefers to use their right hand also uses their left hemisphere *more* than their right hemisphere to name pictures. The number decreased to 73 percent in consistent left-handers, with the more balanced brains falling in the middle at 85 percent.[15]

There are a few things to notice about these results. First, the more strongly right-handed you are, the more *likely* it is that your left hemisphere will be uniquely qualified for the kinds of computations that speech relies on. But as you may remember from the TMS study described earlier, this type of lopsided speech production is also more vulnerable to injury. More balanced brains, on the other hand,[16] tend to be associated with more similar capabilities in the two hemispheres. This means that the right hemispheres of mixed- and left-handers are more capable not only of manual dexterity but also of producing speech. As a result, when either one of their hemispheres gets fatigued with TMS, they experience milder impairments.

Another thing to notice, however, is that the likelihood that your right hemisphere—like Jasmine's—is the talkative one is much lower than 50 percent, even for lefties. This provides a good reminder that

---

15. Note that these statistics indicate which hemisphere was *more* active during speaking, but Knecht also observed that the relevant laterality of speaking, like handedness, lay on a continuum. Some people used primarily one or the other hemisphere, while others used both nearly equally.

16. Pun intended.

most human brains are at least a little bit lopsided, and that differences among us are matters of degree. The fact that speech relies more on the left hemisphere than the right, even in the majority of the *least*-lopsided brains, suggests that the differences in structure between the two hemispheres may be even more important for language, one of the most recently evolved functions of the human brain, than they are for hand control. But the fact that 73 percent of strong left-handers have speech and preferred hand control driven by *different* hemispheres also suggests that their shared computations might not *only* relate to motor control. In fact, there are also pretty remarkable differences in how the two hemispheres *comprehend* language that can provide more clues about how they independently and collaboratively shape the way you understand the world.

The earliest signs of a left-hemisphere dominance for language comprehension can be seen in the auditory cortex—the part of your brain responsible for analyzing sound. A number of studies have shown that the auditory cortex is more active in the left hemisphere than in the right (in *most* people)[17] when study participants listen to speech sounds. In contrast, the right hemisphere is more active than the left when they listen to music! Researchers like David Poeppel and Robert Zatorre have argued that the reason the left hemisphere is tasked with speech comprehension is that it's really good at the computations required to detect *fast* changes in time.[18] Of course, music can be fast too. For a nice link between music and motor control, consider that the fastest drummer ever recorded, Siddharth

17. I couldn't find any research looking at individual differences in this effect in the brain, particularly as it relates to handedness, but based on behavioral research about ear preference, I would assume it follows the same general pattern observed in speech and other lateralized functions.

18. The two researchers have slightly different opinions about what makes the right hemisphere better suited for listening to music. I've included a few relevant papers in the Notes section if you are curious to learn more!

Nagarajan, clocked in at an astonishing rate of 2,109 drumbeats in a minute (though my metronome doesn't go past 250).

But if you want to understand the difference between the "ba" and "pa" sounds in "banana pancakes," your brain needs to be able to detect a *ten-millisecond* timing difference between when someone's vocal cords start vibrating and when their lips part. That's like being able to tell the difference between 5,999 and 6,000 drumbeats per minute, both of which are *way* faster than even the most metal drum solos of all time. Might the left hemisphere's computational advantages be tied to the ability to coordinate or detect things that change very quickly in time?

The short answer is "sort of." If we revisit the "Important Haystack" experiment I discussed in "Introductions," you might also remember that some of the functions that get assigned to one hemisphere or the other unfold over a slower time scale.[19] So what explains whether a particular linguistic function will come to rely on the left or right hemisphere in any particular individual?

One possibility, described by Elkhonon Goldberg and Louis Costa in the early '80s, is that the critical structural difference that drives the specialization of our two hemispheres is how they are *wired*. To be more specific, the two scientists proposed that different patterns of connectivity between hemispheres shape the extent to which the brain areas *within* each hemisphere communicate with one another.[20] According to Goldberg and Costa, the left hemisphere consists of many small, "informationally encapsulated" brain regions. These become the expert "modules" I described at the beginning of the chapter—designed to perform well-specified computations on

---

19. Typical adult reading rates vary between around 200 to 300 words per minute, which is *much slower* than your average drum solo.

20. We'll talk about the mechanisms of such communication in the next two chapters.

specific types of inputs without being influenced by whatever their neighbors are doing. This means that in most typically lopsided brains, the extent to which the left hemisphere contributes to any *function* depends on whether it can be accomplished using a "divide-and-conquer" approach. In language, this might look like moving from sequences of sounds to words, from sequences of words to ideas, and from sequences of ideas to stories.

In contrast, Goldberg and Costa proposed that the right hemisphere's structure, which contains proportionately more connections *between* brain regions, is better suited for jobs that require integration of different types of information into a coherent whole. This explains why, as I mentioned briefly in the assessment section, the right hemisphere tends to get assigned functions like face recognition in most people. To distinguish one face from another, you need to consider subtle differences between many different features, and where they lie in relation to one another. If you don't believe me, try to identify pictures of your friends based on just one of their features, like a nose or a single eye. Without the surrounding features to ground it, this is much harder than you might think.

Let's return to the haystack experiment to see how Goldberg and Costa's ideas might explain our findings about how the two hemispheres contribute to the different ways people understand what they read. As you might recall, the left hemisphere of *all* readers in my experiment was sensitive to the local argument structure of the sentences. This suggests that—at least for people who can read well enough to be in college—specialized processing modules in the left hemisphere are involved in constructing meaning from those sentences based on the linguistic details presented.

The right hemisphere, on the other hand, was differentially involved in skilled and less-skilled readers. In the least-skilled readers, the right hemisphere was sensitive both to the local argument structure *and* to the global, scenario-based context, while the right hemispheres of the most-skilled readers showed no trace of either comprehension process. So what gives?

Our findings are consistent with part of Goldberg and Costa's theory that I haven't discussed yet, which describes how the expert modules of the left hemisphere come to be assigned different functions. According to their theory, complex tasks almost always depend on the right hemisphere initially. Their idea, in a nutshell, is that until you understand the important *parts* of a task, your best bet is to try to use all the information you can to figure out what you're supposed to be doing. When faced with a totally new task, they argue, the right hemisphere's "big picture" or "forest" approach is advantageous. If you've ever tried to get by in a country where you don't know the language or customs well, you might be able to intuit how this works. You can get pretty far using clues like gestures and facial expressions to figure out what you're supposed to do based on the context you're in. But as you gain experience with a new task, you learn which of the smaller parts, or "trees" that make up the forest, are important for the job at hand. And when this happens, your brain develops faster and more effective strategies that rely on specialized processing modules. When it does, it depends less and less on the bigger picture to figure out what's going on.

Consistent with Goldberg and Costa's theory, a number of functions shift to relying more on the left hemisphere with experience. For instance, most infants use both hands equally poorly in their earliest phases of life. It's only around one and a half years old, as they develop increased experience manipulating things, that they begin to show a stable preference for using one hand over the other.[21] The

---

21. This timeline depends on how hand preference is measured (grasping or bimanual manipulation styles), but it is likely later for toddlers with more balanced brains!

same thing happens with language, which initially gets processed by both hemispheres, and shifts to left dominance in most as proficiency increases. Though bilingualism is more complicated,[22] a number of studies also suggest that people's *second* languages rely more on the right hemisphere, particularly if they are learned later in life and the speaker is less proficient than in their first language. A few smaller studies have even shown that if you compare people with significant musical expertise to novices, you find a shift toward left-lateralized music processing in the experts.

To summarize what we've discussed about language in *your* brain so far, there are two things I'd like you to keep in mind: The first is that, according to Goldberg and Costa, one of the key computational differences between the two hemispheres is driven by how they're connected. In the most lopsided brains, the left hemisphere takes more of a divide-and-conquer approach, using expert modules to focus on the tree-level details, while the right hemisphere is specialized for looking at the big picture, or the forest level. One thing that remains unclear, based on the limited research on extreme lefties, however, is whether the right hemispheres of people like Jasmine with reverse laterality end up with expert module wiring. For now, suffice to say that although all brains have both forest and tree capabilities *somewhere*, the more lopsided your brain is, the more likely it is that you will come to focus on specific features, or details, to solve complex problems, while those with more balanced brains are more likely to rely on the big picture.

And keep in mind that for both types of brains, experience with a particular task can shift you toward being a more detail-oriented

---

22. For one thing, there are many different kinds of language experience that can cause someone to be bilingual, and these have different implications for how bilingual brains work. We'll touch on some of my research on bilingual brains in Chapter 3.

information processor. In fact, even functions like hand control are shaped by experience.[23] For example, one study that examined the brains of "forced" right-handers—people who showed an early preference for using their left hands but were forced to use their right hands to conform to societal norms—found that their motor cortex was indistinguishable from "natural"-born right-handers. This just goes to show that nurture, or our experiences, can override the natural predispositions of our brains to a certain extent.[24]

Now, to get a better understanding of the implications of having balanced or lopsided brains, let's take your brain out of the laboratory and get a look at how the functions of your two hemispheres might work "in the wild" to build your understanding of the world you inhabit on a day-to-day basis.

Lopsided functions: The stories your brain tells

So far, this chapter has focused a lot on *mechanical* explanations of how your two hemispheres work. But if you want to get a better sense of how your balanced or lopsided brain shapes the way you think, feel, and behave in the real world, we're going to need to return to the idea of *why* you've got two hemispheres that work differently in the first place. This brings us back to the idea of specialization of *functions*, which, you may remember from the beginning of this chapter, is a very evolutionarily old engineering design. And according to Joseph Dien,

---

23. That whole nature-versus-nurture thing isn't going away, whether you like it or not!

24. However, I do not recommend forcing anyone to be right-handed. It's worth mentioning that a different part of the brain—one that is implicated more generally in control—was *smaller* in forced right-handers. This might be because their left hemisphere had to inhibit, or handicap, their right one in order to overcome their natural predispositions.

the structural differences between hemispheres in the human brain evolved for jobs much older than language.

In essence, Dien proposes that lopsided brains—those that can understand the world in multiple ways at once—have a critical evolutionary advantage. They can focus in two directions at the same time. Much like the model's namesake, the Roman god Janus, is depicted with one face looking forward and another looking behind, Dien proposes that our brains evolved so that one hemisphere (the left)[25] is primarily focused on predicting the future, so that it can help you make the best decisions about what to do next, while the other (the right) is tasked with understanding what's happening *right now*. This idea is related to another function-based distinction of the two hemispheres—one that proposes that the left hemisphere initiates "approach" behaviors, while the right hemisphere focuses on "avoid" behaviors. In either case, the idea is that the kinds of thoughts, feelings, and computations your brain engages in, either to predict the future or to find good things, might come into conflict with the kinds of thoughts, feelings, and behaviors you need to understand the now, or to try to avoid things that might kill you. Also keep in mind that although these theories focus on the functions, or the potential *reasons* your hemispheres might develop structural specialization, they are not incompatible with Goldberg and Costa's descriptions of the structural differences between hemispheres. It's entirely possible that experienced, module-based processors evolved precisely *because* they are useful for executing the fast, specific computations that one needs to predict the future, while integrated, global processors are needed to execute the complex pattern-recognition computations involved in understanding what's happening now, and for

---

25. It might be worth noting that none of these laterality theories attempts to account for individual differences.

relating it to your past experiences to figure out whether something is dangerous or not.

So what might this mean for how balanced and lopsided brains work in the wild? Imagine you were a proficient English-speaker who heard this seemingly straightforward sentence: "They are cooking apples." The meaning of this sentence is *ambiguous*, but I'm willing to bet that most of you didn't feel the slightest bit confused when you read it. This is because your brains have so much experience with sentences like this that your left hemisphere sent the words on the page to be processed in specialized modules. These modules made word-by-word decisions about what the sentence *meant* as it unfolded in time. And unless your brain is faced with evidence to the contrary—say, when there is a mismatch between the context a sentence occurs in and your brain's first interpretation of what it means—it just plugs along, *predicting* the most likely meaning of your sentence based on its past experiences. In fact, I'm willing to bet that most of you interpreted "They are cooking apples" as a sentence that describes some person or people (aka "They") who are executing the action of "cooking," in the present tense, on a series of objects called "apples" without even fathoming what the alternative interpretation might be. But the real question is, would *your* brain have come up with a different understanding of that sentence if it were presented in the following, context-rich environment?

Jasmine walks into the kitchen and finds a bunch of apples in a brown paper bag. They look different from the apples in the fruit bowl. They're very ripe, and some of them are a little bruised. She turns and asks, "What are those for?" pointing to the bag on the counter. "They are cooking apples," I respond.[26]

Boom! Now most of your brains probably built a different inter-

---

26. To be fair, this scenario is pretty implausible if you know how little I cook, but this section is about building narratives, so I took some liberties!

pretation of the sentence. This time, "They" refers to the *apples*, and "cooking" is an adjective that describes what *kind* of apples they are!

I know, language is wild and wonderful.

But the point I really want to make here is that it creates plenty of opportunities for *different* interpretations of the same input. And as I showed in the haystack experiment, the extent to which a person uses context to drive their interpretation of the details of a sentence varies across different brains. So whoever wrote the infamous World War I article titled "French Push Bottles Up German Rear" probably didn't realize that *their* way of understanding that sentence was *not* the only way. I'm willing to bet that they either had a balanced—"forest"—brain or their brain was so steeped in the *context* of battle tactics that it didn't even notice the fact that the word "push" is used much more frequently as a verb than it is as a noun, and that the reverse is true for "bottles."[27]

Whether your brain relies on the forest or the trees to figure out what's going on, the part I find most remarkable about this is that we don't walk around feeling confused *all the time*, despite the fact that we're constantly faced with incomplete or ambiguous information. The reason for this is that your brain simply *fills in the blanks,* using different types of information and computations to figure out what's happening. And as you'll learn throughout this book, this creates ample opportunity for *different* ways of interpreting the same input. Using its different mechanisms for understanding the world, your brain builds a more concrete and complete *story* than it actually has the data to support. And I'm not talking only about how your brain interprets the stories it *reads*. I'm talking about the stories it *creates* as it produces your experience of reality.

In fact, differences in the way the two hemispheres contribute to

---

27. The last chapter in this book is all about the challenges that occur when we assume that other brains interpret the world like ours do.

this storytelling process lie at the heart of the long-held pop psychology myth that the left hemisphere is "analytical," while the right hemisphere is "creative." Though the notion of an analytical/creative distinction is not exactly right, the original idea arose from the observations that Roger Sperry, Joseph Bogen, and Michael Gazzaniga made when studying a fascinating group of patients who have the connections between their hemispheres surgically severed (a procedure known as a callosotomy) to control severe epilepsy. Though the callosotomy surgery prevents seizures from "spreading" from one side of the brain to the other, it also largely prevents the two hemispheres of these patients from sharing information with each other. And this provides a rare but powerful opportunity to ask what each hemisphere *knows*, without any input from the other one.

The researchers used the same trick I used in my reading study. They flashed words or pictures to one side of a screen or the other, leveraging the way visual information gets sent to the opposite hemisphere. Consistent with the idea of a left-hemisphere dominance for speech, *most* callosotomy patients can talk only about the words or pictures that are presented on the right side of their screen, which are seen by their left hemispheres. But here's where things start to get interesting. If you present a picture on the left side of the screen, so it is only seen by the right hemisphere, and then you ask a callosotomy patient what they saw, they will typically answer, "I didn't see anything." This is because their left hemisphere is speaking, and it *didn't* see anything. *But* if you give their left hand a pencil and ask the right hemisphere to *draw* what it saw, it can! How wild is that? But wait, this story gets even weirder.

In the process of conducting these investigations, Michael Gazzaniga, who began this research as a graduate student working with Sperry, noticed something fascinating: Sometimes, when a patient saw what their left hand was doing (with both hemispheres), they would spontaneously make up a story to help bridge the inconsistency between what they said they saw (nothing) and what they drew. In one example that Gazzaniga captured on video, two different pictures were flashed on the screen at the same time: A sun was presented on the right side of the screen, and an hourglass timer on the left. "What did you see?" he asks the patient. "The sun," the patient responds, based on the information that is available to his left, speaking hemisphere. "Can you draw it?" Gazzaniga asks, putting a pencil in the patient's left hand. He draws an hourglass, because that is what his right hemisphere, which controls his left hand, has seen. Now the speaking left hemisphere can see what his left hand has drawn, and it starts to make up a story about *why* it did that. "What did you see?" Gazzaniga asks again. "A sun," the patient responds. "But I drew a timer *because* I was thinking about a sundial," the patient says, *confabulating* a plausible connection between what his left hemisphere knows and what it observed. And voilà, the patient's brain was caught red-handed in its storytelling process.[28]

Over the course of these experiments, Gazzaniga inadvertently discovered that the speaking part of our brain also seems to generate *causal explanations* that relate events to one another. Since then, Gazzaniga and other scientists, including myself, have gone on to experimentally investigate differences in these "inferential" processes in both hemispheres of intact participants and callosotomy patients alike. The general consensus is that in most people, the left hemisphere generates hypotheses about what might link two events together based on details it deems relevant. And because of its ability to do so, Gazzaniga named the left hemisphere "the interpreter." From that observation, subsequent journalists and researchers created the idea of an *analytical* hemisphere.

---

28. I've added a link in the notes to an interview you can watch on YouTube between Alan Alda and Gazzaniga, featuring callosotomy patient "Joe." It's fascinating!

As you might suspect, this specific kind of analysis—the kind that reverse-engineers causality from observed events, is really important for predicting the future. And let me make it quite clear that your healthy brain does this all the time. Much like the left hemisphere of a split-brain patient makes up stories to fill in the gaps when it observes a behavior it isn't controlling, your brain is constantly creating a personal narrative that weaves a causal explanation through the actions it observes you doing. Though your two hemispheres are probably well connected, your interpreter still needs to fill in the gaps about why you do the vast majority of things that are, in all actuality, driven by subconscious brain processes.[29] But this happens so frequently, and so fluidly, that we are largely totally unaware of our brain's storytelling process.

If you're still not convinced that *your* brain is making stuff up, try to remember a time when you woke up after a lapse in consciousness. The most striking memory I have of such an event happened during graduate school.[30] The shortest version of my story begins with me "waking up" with my head sticking out the front door of my apartment. My first conscious experience was hearing my inner voice saying something like "I must have taken a nap."[31] Almost immediately afterward, my brain ran a reality check on that interpretation. The idea was followed by something like "I must have been really tired to have taken a nap in the doorway!" And then, when faced with conflicting information like "Wait—I don't take naps in doorways!" my brain started searching its memory bank for different solutions. When it did, it was able to pull up the memory of burning skin and talking to a nurse on the phone, and my left hemisphere was able to

29. The details of this will come up again in the "Navigate" chapter.
30. Don't worry, this story is totally PG.
31. Being overly tired and succumbing to undesired naps in weird places was pretty "normal" for me as a single mom in graduate school.

use that new data to construct the new, more plausible, story that I had fainted![32]

Although this process becomes noticeable in unusual situations like mine, even sleep can produce a lapse of consciousness that lets you catch your interpreter in the act. You're more likely to notice when things are unexpected—like when you wake up in a different place than your bedroom, and your sleepy brain has to process why the things you're seeing and hearing are *not* the things it expected. When this happens, you can sometimes "hear" your interpreter trying to figure out where the heck you are. In those fleeting moments, where the dots aren't easily connected, you might become more aware of your brain's storytelling processes. And although this has mostly been studied in the domain of reading, there are hints that people with more balanced brains rely more on the broader context to interpret what is going on, while more lopsided brains may focus in on the individual details first.

I completely understand that this might make you feel *weird*, but trust me—if your brain didn't tell you stories, you'd be in big trouble. At the pace that natural conversations occur, for example, if you took even five seconds to wonder what *else* a person might mean when they say "They are cooking apples," you'd miss the next *ten words* they said. Good luck trying to figure out what's going on after that!

One thing I've always wondered about this, which hasn't yet been studied systematically, is how tightly related our conscious storytelling processes are to speech. According to a tweet that went viral in January 2020, the experience I described, in which you can "hear" your thoughts expressed verbally in your mind, is not universal. In fact, a significant number of people (including my husband,

---

32. I had a weird reaction to some vitamin supplement I took—a niacin flush? Do not recommend.

Andrea) don't experience their internal thoughts as words at all. This makes me wonder whether, in some balanced brains, the functions of speaking and interpreting get assigned to different hemispheres. If that happens, does it fundamentally change the nature of your "personal narrative"? What might the nonnarrative version of this narrative be?[33]

We have hints that come from observing patients whose two hemispheres have been severed. Many of them seem to "identify" more with whatever is going on in their left hemisphere than they do with their right. In the days that followed her callosotomy surgery, a patient named Vicki described her frustrating experience with everyday tasks, like grocery shopping or picking out her clothes for the day. "I'd reach with my right [hand] for the thing I wanted, but the left would come in and they'd kind of fight . . . almost like repelling magnets." In these anecdotes, two things are clear. First, when the two hemispheres are surgically separated from each other, they can come up with different *ideas* about how to behave, based on their unique ways of understanding the world. The second is that the subjective experience these patients *report verbally* is consistent with how their left hemispheres behave.

Fortunately for most of us, approximately 150 million high-speed neurons, collectively referred to as the corpus callosum, connect our two hemispheres, allowing them to rapidly share their views of the world with each other. And so, while we may still have difficulty deciding what to wear, we experience these decisions from the perspective of a single, unified "self" that has access to the integrated outputs of both hemispheres. The principles of neural engineering that allow us to control the flow of information from one brain area to the next will be the focus of the next two chapters.

---

33. According to Andrea, the nonverbal types have "consciousness Netflix" (on mute), while the verbal types have "consciousness podcasts."

Summary: Different degrees of forest- and tree-level brain computations shape our understanding

Before we move on, let's take a minute to review some of the key concepts that were covered in this chapter. There were a lot of them, and the coming chapters will build on them further to give you a better idea of how your brain works. One critical topic we discussed is the relation between the structure of your two hemispheres, and the different computations they perform. In *most* people, the left hemisphere seems to be optimally structured for a divide-and-conquer approach—one that uses modules to execute specialized computations that don't interact with one another. This is kind of like building an understanding of a forest one tree at a time. The right hemisphere, on the other hand, takes a big-picture approach, integrating as much information as possible from different processing centers into a coherent story about the event or scenario unfolding around it. This is kind of like saying, "I know I'm in the forest, so that vertical thing in front of me must be a tree!"

Though these structural asymmetries may be reversed in a very small percentage of the population, the biggest difference between brains is the degree of specialization between hemispheres. And though there are clearly advantages that pushed us to have two hemispheres that can see the world from two different perspectives at one time, the disadvantages of having extremely lopsided brains include increased vulnerability to injury, as well as potentially weaker functioning in things that require seeing the big picture.

We also discussed the fact that the assignment of a particular *function*, like speech or sentence reading, to one hemisphere or the other depends not only on how different the two hemispheres are from each other but also on how much experience a person has performing a particular task. But there are even subtler differences we haven't talked about yet that can dynamically influence how much one or the other of your hemispheres contributes to any function.

For instance, one remarkable study by Casagrande and Bertini measured patterns of brain activity and relative hand skill in a small group of 16 healthy, right-handed volunteers during various parts of their wake/sleep cycles. They showed that *all* participants had greater activity in the left hemisphere and better skill in their right hands during the waking hours, but immediately before falling asleep, and immediately after waking up, their right hemispheres were more active, and their left hands were more skilled! This means that each of us has an opportunity at the beginning and end of the day to catch a glimpse of what the other half of our brains might be "thinking," although we may not be as adept at "talking" about it.

And if you think that's bizarre, multiple experiments have demonstrated that something as simple as clenching one of your hands into a fist for a prolonged amount of time can influence your pattern of thinking, feeling, and behaving by changing activation levels in one of your hemispheres. For instance, some of these studies have shown that if you clench your left hand (activating your right motor cortex), you can increase the relative degree of "avoid" feelings, or amount of dislike you report for any given stimulus, while clenching your right hand and activating your left motor cortex increases your "approach" motivation, or the degree to which you report liking something. These research findings remind us that while some proportion of our differences in lopsidedness is relatively stable, changes do occur *within us*, both slowly as we become more experienced with certain processes across our life span, and more quickly as we move through various states of wakefulness or respond to different environmental factors that activate one hemisphere more than the other. And so, if you *feel* like a different person when you wake up in the morning or go to sleep at night, it might help to know that there are fundamental differences at play in how your brain works. In the next chapter, we'll turn our discussion to some more nuanced aspects of your brain design, discussing the way its chemical ingredients shape the types of information it shares, both within and between hemispheres.

CHAPTER 2

MIXOLOGY

The Chemical Languages of the Brain

In this chapter, we're going to zoom in to discuss the smallest of your brain's design features—your neurotransmitters. To put it simply, your neurotransmitters are the chemicals that your neurons rely on to communicate with one another. Though all brains use them, the human brain has *hundreds*[1] of different kinds at its disposal. And at any given point in time, *your* brain is floating in a cocktail made up of its own unique mixture of these ingredients.

If you've ever smoked marijuana with a group of friends,[2] or enjoyed an alcoholic beverage or three at a social event, you probably already understand some of the most important things about your brain's mixology. The first is that substances that alter your brain chemistry can change the way you think, feel, and behave—sometimes in dramatic ways. And the second is that these changes don't look the same in everyone. Both phenomena relate to the ways

---

1. This number varies, depending on whether you consider classes of neurotransmitters or individual chemical compounds. For the purposes of this book, it probably doesn't matter, since the vast majority of the research has been conducted on only a handful of them.
2. And inhaled . . . hey, it's even legal in lots of places now!

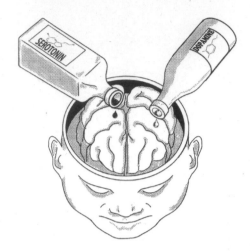

different brains use chemicals in their communication systems. In this chapter, you'll learn *why* the littlest parts of you can make such a big difference in how you work!

Take caffeine, the most popular drug in the world,[3] as an example. Drinking a cup of coffee, tea, or other caffeinated beverage affects your chemical mix in multiple ways. The most wonderful of these, in my opinion, is that it increases the availability of a neurotransmitter called *dopamine* in your brain. Dopamine is one of the most important ingredients in your neural cocktail, because it is the chemical the pleasure circuits in your brain use to communicate. And because all brains are motivated to feel good, your brain's dopamine circuits are heavily involved in learning and decision-making. Their goal is to shape the decisions, big and small, that move you through the world in a way that earns you the most pleasure points. This makes the popularity of caffeinated beverages a no-brainer.

And now imagine the possibility that the differences in baseline levels of dopamine between two brains might exceed the difference

---

3. According to a survey conducted in 2014, 85 percent of the U.S. population consumes at least one caffeinated beverage per day . . . and that doesn't even count chocolate!

you feel before and after your morning cup of caffeine. One person's baseline mental state could feel like you after a shot of espresso, while for another, your morning brain *before* coffee or tea might be their personal high.

To get a better idea about how the levels of the different ingredients in your neural cocktail influence how you think, feel, and behave, let's take a closer look at the relation between structure, computation, and function that we began discussing in the last chapter. For starters, consider the fact that the computational differences we talked about in "Lopsided" arise because of the way that networks of *millions*, or even *hundreds of millions* of neurons work together. But as we touched on in "Introductions," when it comes to individual neurons, they all perform *essentially* the same job. Their computations involve listening to the "gossip" of other neurons around them and deciding whether they have enough evidence to pass their own signal down the line.

In fact, the extent to which any individual neuron will contribute to some *function*, like language, depends largely on *where* it is in the brain. This is because location is one of the main factors that determine *what* gossip a neuron is listening to. In other words, the function a neuron performs depends almost entirely on the inputs it performs its computations on.

In 1988, a group of neuroscientists made this point very clearly by surgically rewiring a newborn ferret brain so that the neurons carrying signals from its eyes were connected to the neurons that usually process information coming in from the ears. When they did, they created a ferret that learned to "see" using its auditory cortex—the part of the brain usually assigned the job of hearing. Eventually, the auditory cortex in the ferret was able to take over the *function* of the seeing when it was fed the same inputs.[4]

---

4. It might be worth noting that these rewired ferrets didn't see as well as ferrets

But we can also observe fascinating naturally occurring evidence of what happens when neural signals get crossed, in humans who exhibit synesthesia. This condition, which occurs in an estimated 2 to 4 percent of the population, involves the merging of two *unrelated* streams of sensory information in the mind and brain. The outcome can range from people who experience some sense of shape (for example, square or round) when presented with different food tastes to the more frequent case of seeing colors when presented with particular letters or words. The moral of the story is, when you've got a thundering 86 billion neurons gossiping in your brain, it needs systems for organizing who is talking (and listening) to whom.

True to the spirit of this book, this engineering problem—the need to keep track of overlapping signals between neurons to organize their functions—creates a brain-design space that has several different solutions. Ironically, this relatively large problem space occurs in a very tiny physical space—in the *synapse* between neurons—a 0.02-micron gap, 1/2,000th the diameter of a piece of hair, that separates them from one another. It's here that your mixology plays a pivotal role in shaping the function of your neurons, by determining how well they can communicate with others.

To get a handle on how this works, let's use as a model for neural communication the game called Telephone that we used to play when I was a kid. In Telephone, one person comes up with some secret message and then whispers it to the child next to them. That child then whispers it to the next child in the chain, and the process is repeated until the message gets passed all the way around a circle back to the sender. The funny part of the game is that when your original message comes back to you, it's usually been changed into

---

that hadn't been mucked with. This goes to show that, as we discussed in the last chapter, there are reasons that nature assigns certain functions to certain regions of the brain. The auditory cortex was capable of doing the job of the visual cortex but was not as capable of doing the job as the visual cortex was.

something totally different. At each transition between people, the conditions caused by the combination of a soft signal (a whisper) and a noisy room (usually full of giggling children) drive some level of interpretation or improvising on the part of the listener. The result is that a message like "Do you want banana pancakes?" can easily be transformed into something like "Cthulhu haunts bandana man-scapes."

Though it might be hard to believe, your brain works something like this: In its version of Telephone, the whispering between neurons is accomplished by the release of neurotransmitters. Much like a message whispered between humans temporarily assumes the physical form of a sound wave traveling from mouth to ear, the messages exchanged in the space between neurons must temporarily take the physical form of chemical packages. And this is where your brain's smallest design features, the chemical ingredients of you, start to shape the way you work.

For starters, each neuron's ability to whisper, or send messages, to other neurons in your brain is *finite*. To accomplish their goal, they need access to their preferred chemical ingredient or ingredients. In fact, if a neuron gets really excited about the gossip they're hearing, they can release *all* of their chemical messages into your neural cocktail and become temporarily mute. It's kind of like running out of "swipe-rights" on Tinder.[5] The blind spot you see after looking at the sun, or a camera flash, is a real-world example of this.[6] Such an intensely bright light excites the hell out of the neurons in the back of your eye, causing them to release all of their chemical packages while trying to tell their social networks about it. Since looking at the sun

---

5. I know, some of you didn't even know it was possible to run out of swipe-rights on Tinder—and that, my friend, is probably directly related to the ingredients of your neural cocktail.

6. Just an FYI, don't do this. Looking at the sun really causes permanent damage to your retina. It's not an urban legend.

is not good for you, let's demonstrate the effect with a safe at-home experiment.

Focus your eyes on the center of the vinyl-record-shaped image below for ten seconds. Then move your eyes to a blank spot on the page next to it, or move them around in the world, or close them. Feel free to experiment however you'd like with the safe, and temporary, *hallucination* that follows.

What you should see is an inverted "afterimage"—a bright disk with a darker circle in the center that looks kind of like the Eye of Sauron. This happens because the neurons whose job it is to give you the experience of seeing a bright light in the center of your visual field, and those whose job it is to detect darkness in the ring around it, ran out of neurotransmitters.[7] And when those neurons went radio silent, the connected neurons whose job it is to "listen" and pass along their own opinion of the state of things interpreted that silence

---

7. In the real world, we move our eyes multiple times every second. This lets us take in information in little bits, and also allows our visual neurons to replenish their neurotransmitter supplies.

as evidence that the *opposite* conditions must be occurring in the world.

The resulting hallucination allows you to experience, firsthand, what can happen when your brain *interprets* the incomplete information it receives from the world "out there" against the noisy background that is a fundamental part of its information processing. It's also a salient, and drug-free, way to remind yourself that *your* experience of reality is created by your brain.

But every brain also has different amounts of the ingredients they use to communicate to begin with. For example, if we were to do an experiment about afterimages in the lab, we would likely find differences in how long each participant needed to stare at the disk before seeing the afterimage, and differences in how long their afterimages persist. In fact, a series of experiments by Richard Atkinson, former president of the University of California college system and director of the National Science Foundation, found that individual differences in the length of time afterimages are perceived relates to hypnotic susceptibility—which has also been linked to individual differences in neurochemistry.

To get a better understanding of how differences in the chemical ingredients of your neural cocktail influence your characteristic ways of thinking, feeling, and behaving, we'll need to get into more of the details of this design feature. In the next section, I'll explain the mechanisms that operate on our chemical communication systems, to highlight the costs and benefits of different design choices.

The costs and benefits of differing neurotransmitter levels

It's pretty easy to imagine what the costs of not having enough neurotransmitters may be. As the afterimage experience demonstrates, when your neurons "run out of gas," they stop passing along their signals in your brain's game of Telephone. And when some of your neurons go mute, it fundamentally changes the way you experience the world. In fact, if you're one of the estimated 7.8 percent of adults in the United States who suffer from depression, you have experienced firsthand how consequential it can be when your brain runs low on feel-good dopamine or related neurochemicals.

So why doesn't the brain just make sure every neuron has an infinite supply of its preferred chemical communication system? The most obvious answer is that even the smallest ingredients take up room. And we know space is limited. But the truth about the costs and benefits of soaking your brain in all the chemicals—feel-good or otherwise—is more nuanced than that.

To better understand the costs of having *too much* of an ingredient in your neural cocktail, we need to get a bit more specific about how your neurons use these chemicals to communicate with one another. The first thing to note is that when a sending neuron "whispers" its chemical message out to the neighboring neurons, it has no way of *controlling* who will receive the packages it sends. Instead, it drops its chemical signal into the cocktail rather indiscriminately. This creates a situation that's dramatically different from the children's game of Telephone, where messages are passed directly from sender to receiver in a one-to-one fashion. In your brain, any listening neuron might be privy to the messages coming from ten thousand whispering neurons at once!

One problem with this brain design is that it creates a lot of noise.[8] And the more chemical messages there are floating around in the background, the more difficult it is for any listening neuron to

---

8. As it turns out, this noise is both a feature and a bug, as it drives the interpretation processes that are critical to your brain's computations.

detect a whispered signal from its neighbor. Additionally, in a perfect world, each neuron's chemical message should correspond to some time-locked event in either your outer or inner world. But if that message isn't *received* immediately, it can continue to echo around in your brain. And as the time increases between when a message is sent and when it's received, so does the chance that the message is no longer relevant. As you might imagine, this creates a completely different type of noise. Imagine if some neuron in your brain was trying to decide what to do, based on some combination of what's happening around you *now* and what was happening five minutes ago. For most of the things you do on a moment-to-moment basis, this would be a disaster. So while not having enough of a particular ingredient can force parts of your brain to go mute, having too much can cause "gossipy" conditions in the brain in which messages received by the wrong neurons, or by the right neurons at the wrong time, can wreak havoc on your understanding of the world around you.

Fortunately, your brain's chemical communication process isn't as willy-nilly as I made it sound. For one thing, proximity plays a big role. When your nearest neighbor is only a fraction of a hair's width away, they're much more likely to receive your message than someone trying to listen from a different neighborhood. And if your brain thinks it's important for neuron B to be listening more to neuron A than to its other neighbors, it has mechanisms that allow neuron B to grow more "ears" or *receptors* that are close to neuron A. And voilà, *this* is how you learn!

Also, to keep it from getting *so loud* in your brain that your neurons can't hear their intended messages, your brain has two different ways to turn down the volume. The first is *reuptake*—a recycling-like process that allows sending neurons to reabsorb any undelivered chemical messages they find and reuse them. While neurons with effective reuptake processes can make a little bit of their ingredient go a long way, those on the listening end of these neurons have a very limited window to receive their messages before they get stamped Return to Sender, so to speak. The second method for turning down the volume is metabolic. Among the critical ingredients of everyone's neural cocktail are enzymes that break down the neurotransmitters they encounter like Pac-Man gobbling pellets. When they're done, the remains of the whispered message are uninterpretable. But some of these message bits can also be reabsorbed by sending neurons and used to rebuild new, fully formed neurotransmitters. Taken together, these *four* design features—the availability of neurotransmitters in sending neurons; the effectiveness of recycling, or reuptake, mechanisms in sending neurons; the number and distance of receptors (or ears) in receiving neurons; and the amount of enzymes working to break down unsent messages between neurons—jointly influence the amount of any ingredient in your neural cocktail.

But there's one more design feature that will allow us to get an idea of the levels of some key chemical ingredients in *your* brain. One of the critical ways your brain manages its elaborate and somewhat chaotic game of Telephone is that not all neurons "speak" the same chemical language. And though some neurons can send messages in bilingual formats, each receptor of a receiving neuron can respond only to one specific language.[9] Critically for figuring out your mixology, groups of neurons performing the same job tend to organize themselves based on the chemical language they speak. As you'll learn in this chapter, the availability of certain types of chemicals can be traced to specific brain functions. And this is how we'll start the reverse-engineering process to figure out what's in your mix!

---

9. This is because each neuroreceptor is a protein that takes on a particular shape, allowing it to fit together with a specific neurotransmitter like a lock and a key.

Assessing neurochemistry through personality characteristics

One way to reverse-engineer your mixology is to start with a list of the characteristic patterns of thinking, feeling, and behaving that describe you, on average. To help streamline the process, I've included a list of adjectives that I modified from Gerard Saucier's "Mini-Markers" personality test. As you'll learn in this chapter, many of these characteristics have been related to individual differences in neurochemistry.[10] Your job is to think about each word and how well it describes you *on average*,[11] compared to other people you know in your age group. Then give yourself a score between –3 (extremely inaccurate) and +3 (extremely accurate) based on how well you think each adjective describes you. For example, though I am not *nearly* as energetic as I was when I was twenty, I do tend to have more energy than many of the people I know in my age group, so I would probably give myself a +2 next to the characterization "energetic."

Of course, the more honest you are with yourself, the better your estimate will be. If you are unsure, or just feeling particularly brave or curious, you can also ask someone close to you to fill out the questionnaire and compare notes.[12] There are some characteristics, like "cooperative" and "kind," that have positive connotations, and others, like "disorganized" and "selfish," that have a more negative vibe.

---

10. There are *many* different studies linking individual personality and temperament to neurochemistry, some of which we'll discuss in detail in this chapter. But if you'd like to read some contemporary scientific reviews, see the review papers by Depue and Trofimova that I listed in the Notes section.

11. Obviously, these things can fluctuate on a day-to-day basis, so you'll have to try to average across those fluctuations and find your own "typical" level.

12. As a heads-up, if you and the person close to you don't see eye to eye, it's totally OK! Trying to figure yourself, or someone else, out is really tricky. We'll get into *why* at the end of the book. For now, I hope you've got the right chemical ingredients on board to trust me when I say you're not alone!

This can lead to a social-desirability bias, or the tendency to rate yourself higher on positive characteristics and lower on negative ones. Many personality assessments have trick questions built in to adjust for this, but I didn't do that. Remember, my goal is to help you get to know yourself—and this is just the next step! Finally, this is not a vocabulary test. So if you find yourself unsure of what a word means, look it up. And if you start worrying about nuances or multiple meanings of the word, you're probably thinking too hard.

CHARACTERIZATIONS ASSESSMENT

Next to each adjective, write the number on the scale from –3 to +3 that summarizes how well it describes your typical way of thinking, feeling, or behaving compared to others your age:

–3	–2	–1	0	1	2	3
INACCURATE						ACCURATE
Strongly	Moderately	Mildly	Neutral	Mildly	Moderately	Strongly

1. Anxious ____	11. Intellectual ____	21. Rude ____
2. Bold ____	12. Jealous ____	22. Shy ____
3. Calm ____	13. Kind ____	23. Selfish ____
4. Cold ____	14. Moody ____	24. Systematic ____
5. Cooperative ____	15. Nervous ____	25. Talkative ____
6. Creative ____	16. Outgoing ____	26. Timid ____
7. Disorganized ____	17. Philosophical ____	27. Unenvious ____
8. Efficient ____	18. Practical ____	28. Unyielding ____
9. Energetic ____	19. Quiet ____	29. Withdrawn ____
10. Imaginative ____	20. Relaxed ____	30. Worried ____

We're going to use these scores to see where you fall on two different dimensions of personality. But first, I'd like to unpack the science of personality a bit. One thing to note is that the extensive body of personality research that's been conducted on *hundreds of thousands* of individuals has shown that some of these characteristics cluster together. Take "anxious" and "jealous" as examples. These adjectives describe two different feelings: *anxiety*—the state of being worried, fearful, or having an uneasy mind, and *jealousy*—a negative feeling caused by wanting what someone else has, or by believing that a love interest likes, or is liked by, another. If I asked you to think of a time that you felt anxious but not jealous, or vice versa, you could probably come up with one pretty easily. However, people who rate themselves as *consistently* feeling more anxious than others are also likely to rate themselves as being more jealous than others, on average. And the reverse is also true—people who rate themselves as feeling less anxious than others are also likely to rate themselves as feeling less jealous than others, on average. This suggests that there is some more fundamental *factor* that varies between people that influences both mental states. And though the experts continue to argue about how *many* of these factors are needed to describe the differences in the ways people think, feel, and behave, they largely agree that the factors relate to differences in our neurobiology.[13] The two dimensions I've chosen are those that have been most extensively linked to individual differences in neurochemistry specifically.

---

13. Two of the most prominent theories of the biological basis of personality were put forth by Hans Eysenck and Jeffrey Gray. Eysenck's theory is based on three fundamental dimensions: Extraversion, Neuroticism, and Psychoticism. Gray's theory is based on two dimensions: Anxiety and Impulsivity. If you are interested in learning more, I've included a paper by Matthews in the Notes that does a great job comparing the two.

To see where you fall on the first dimension, first add up your scores on the four adjectives that are positively associated with this factor: bold, energetic, outgoing, and talkative.

Now add up your scores on the following four items, which are associated with the *negative* version of this factor: quiet, shy, timid, and withdrawn. When you're done adding them, reverse the sign of this second number, since these characteristics move in the opposite direction to the others. For example, if you *strongly agreed* that quiet, shy, timid, and withdrawn characterize you, you should end up with a score near –12. However, if you strongly disagreed with these characteristics, your score should be near +12. Now add your two scores together and divide by eight to get the average of your score on Personality Dimension 1.[14] As a sanity check, this score should now be back to a scale from –3 to +3.

And now let's calculate your score on the second personality dimension. The formula is about the same. First add up your scores on the three adjectives that have been positively associated with this factor: calm, relaxed, and unenvious. Now add up your scores on the following five items that are associated with the *negative* version of this factor: anxious, jealous, moody, nervous, and worried. Once again, reverse the sign of this second number, since these are negatively associated with the factor. Now add your two scores together and divide by eight again to get the average of your score on Personality Dimension 2. And voilà, we've got the first two clues to your mixology! Now let's get started figuring out what they mean!

---

14. Note that this is a quick-and-dirty way to get a sense of your personality. Many professional tests do more sophisticated types of analyses to figure out where you fall along their multiple personality dimensions.

How typical are you?

Before we get into the nitty-gritty details about how differences in neurochemistry relate to personality traits, I'd like you to understand what your assessment might tell you about how typical your neurochemical design features are. Among other things, this might give you an idea about how likely neurochemical studies that don't measure individual differences are to reflect the way your brain works. Personality traits tend to be "normally distributed"—a statistical term used to describe the way typicality changes across different values of some variable. Put simply, if a variable is normally distributed, most people will have values right around its middle (which is 0, in this case). Then, as you move away from that average in either direction, the number of people who have that score drops off pretty fast. Graphic depictions of this are often called bell curves because of the shape they take. Based on this idea, I'd expect that the majority of you (approximately 68 to 70 percent) would land right in the middle of each dimension, scoring somewhere between −1 and + 1. Then, as you move out, the second-largest group of you (25 to 27 percent) would score between −1 and −2 or between +1 and +2. And finally, at the extreme ends of the scale, I'd expect to find 4 to 6 percent of you who scored either greater than +2 or less than −2. The closer you are to the extreme ends of the distribution, the more likely it is that your brains have unusually high or low levels of the ingredients we'll discuss next.

The pleasure principle: How dopamine rewards drive extraverted behaviors

Those of you who scored in the high positive range on Dimension 1 likely identify as *extraverts*, the name commonly assigned to the cluster of personality characteristics associated with "turning outward," or seeking mental stimulation from external things. If you scored lower on Dimension 1, on the other hand, you probably consider yourself to be an *introvert*, the name given to those who "turn inward" and prefer to be engaged by their own thoughts and feelings than by the outer world. However, as we discussed in the last section, most of you will have values nearer to the middle, exhibiting a more balanced profile of turning to your outer or inner world for stimulation. But no matter where you fall on this scale, converging research suggests that your dopamine communication systems are at least partially responsible for putting you there. To understand why, we'll need to discuss the shared goal that unites the neurons in your brain that use feel-good dopamine to communicate: *Motivation for Rewards.*

Earlier in the chapter, I introduced you to dopamine, your brain's pleasure ingredient, via the effects of caffeine. But this comparison isn't perfect, because caffeine has other stimulating effects on your nervous system that aren't related to dopamine. Also, if you are one of the 85 percent of Americans who drink caffeine daily, it probably only raises your dopamine levels a little, for reasons we'll get into in a bit. For now, allow me to walk you through a more complete description of the dopamine pleasure response.

Imagine the following scenario. You're a contestant on the new (and wildly popular) game show called *The Brain Wants What It Wants.* The game involves choosing which of two options will release the most dopamine in your brain. As long as you pick the right one, you'll win that prize! Behind door number 1 is an all-expense-paid vacation to a spa and wellness center in an idyllic mountain location with perfect weather. But behind door number 2 are backstage passes and VIP seating at Coachella.[15] How would you choose?

If you want to win the game, all you need to do is pick the prize you would most *like* to win. The reason is simple, though the processes that drive it are a bit more complicated. Because your brain *wants* you to feel good, it motivates you to make decisions that it believes will be the most rewarding. The more dopamine your brain expects, the more strongly it will drive you to want something.

But wait, we started this dopamine journey by talking about a drug that increases the amount of dopamine in your brain, and now we're talking about spas versus Coachella? The fact is, *everything* that makes you feel good, from meditation to methamphetamine, increases the levels of dopamine in your brain—at least temporarily.[16] Whether you're having a visceral pleasure experience or feeling a more ethereal sense of bliss, if it feels good, there's dopamine involved.

At the end of the day, dopamine is the "point system" that your brain uses to adjust every possible life outcome's *reward value*. In other words, if your life were a video game, the amount of dopamine in your mix would be the way your brain decides whether or not you're #winning. What's neat is that if we take the drugs that directly influence dopamine out of the mix, there's no way *I* can tell, without measuring your neurochemical responses to different things, what

15. And Beyoncé is headlining—again!

16. Of course, the mechanisms by which meditation and methamphetamine increase your brain's dopamine levels, and the associated health benefits or risks, are quite different. For this reason, I would recommend the former and not the latter. But the point I'm trying to make here is that the extent to which a person would find either thing *pleasurable* relates to the size of their brain's dopamine response.

your brain's point system is. This index of "the relative goodness of things" at any given point in time is unique to you. For instance, on a hot day, I like a nice glass of iced tea with lemon better than water or a soda, but not nearly as much as vanilla ice cream. A sincere compliment is *slightly* above vanilla ice cream, though it kind of depends on the source.[17]

But what does this have to do with how extraverted or introverted you are? Of course, extraverts and introverts will systematically put different reward values on certain activities, particularly when they involve interacting with others or seeking external stimulation. And these different values will motivate them to make different choices. But there are also dopamine-related reasons *why* extraverts and introverts want different things.

To understand this, we'll need to talk about how your brain uses dopamine, like a carrot dangling in front of your face, to motivate you. To make a long story short, your brain is playing *The Brain Wants What It Wants* in every waking moment. The main difference is that you don't usually know what's behind each door, or what the outcome of your choice will be, because your brain can't see the future. It can only make its best guess about whether door number 1 or 2 will bring more pleasure based on your previous experiences.[18]

Here's how this works in a nutshell—when your brain finds something surprisingly good behind one of the "doors" in life, it releases dopamine. Not only does this make you *feel good*, it also creates the conditions inside your brain that promote learning. To help you find *more* rewards in the future, dopamine signals increase plasticity so that your brain can grow and change in ways that increase the likelihood that the group of neurons involved in that decision will be able

---

17. Funny animal memes and good sex beat these things handily—but enough about me already . . .

18. We'll talk about this a lot more in the "Navigate" chapter.

to communicate better in the future. The result is that if you happen upon that door again, even if it's been a while and you don't remember what happened last time, your brain will help you *want* to open it again.

Of course, in the reality TV version of *The Brain Wants What It Wants*, we often have to open a lot of doors before we find something rewarding. Imagine that one hot day, you go for a walk in a new neighborhood. You come to a fork in the road and decide, completely randomly, since you've never been there before, to turn left. Much to your pleasant surprise, thirty yards ahead you find a roadside shop that is selling <insert your favorite warm-weather snack here>! Now your brain knows what's coming, because you've been in a shop that sells <favorite snack> before. You know you can buy some, and that it will be delicious! So your brain kicks in with the *wanting*. "Open the door to the shop," it says. "Then stand in line," it nudges you. "Now use your words to ask for what you want, and take out your preferred payment method!"

Though you might take them for granted, each of these tiny actions that move you closer to a reward was shaped by your brain's dopamine reward circuits. And as you bite into your yummy treat, the dopamine rains down on your brain, creating the signal that allows it to strengthen communication between the neurons that were involved in your string of good decisions. At the biological level, these changes amount to increasing or decreasing the connectivity between neurons in your network. And at the behavioral level, these changes increase the likelihood that you will make choices that lead you to rewards in the future.

But there's one more detail that you need to understand in order to realize how such a system could lead to extraversion or introversion. The *more* dopamine your brain releases with any outcome, the stronger this learning effect will be in your brain. So if the shop had been selling <insert your second-favorite snack> instead, the likelihood that you would turn *left* again at the same fork in the road—or even decide to go for a walk in that neighborhood again at all—would be slightly lower.

And here's where the differences between introverts and extraverts come into play. As it turns out, when *unexpected* rewards happen to extraverts, their brains release *more dopamine* than when something with the same value happens to introverts. Mike Cohen, a friend of mine from graduate school who describes himself as "an introvert with periods of time peppered with extraversion," along with his graduate adviser, Charan Ranganath,[19] and collaborators, were among the first to demonstrate this in the lab. To do so, they used a tool many neuroscientists, including myself, rely on to figure out how human brains work—magnetic resonance imaging, or MRI for short. I'll spare you the details of the physics,[20] but those of you who have had an MRI for health reasons know that it is a tool that allows you to get a *very* detailed, three-dimensional image of different tissue types in the body. And while we can use "regular" MRI to get such detailed images of your brain's *structure*, there's an even cooler method, invented just over thirty years ago, that lets us watch it *work*.

To put it simply, when your metabolically expensive brain starts working, your body fuels it by increasing its supply of oxygenated blood. As it turns out, the MRI machine is so sensitive to the properties of different tissue types, it can measure *how much* oxygen is in the blood. And because your brain is *so expensive*, the oxygenated blood gets delivered on an as-needed basis, directed to the specific

---

19. I mention Charan as a coauthor here because he'll return as one of the prominent researchers in the field of curiosity later in the book.

20. This makes me feel a little bit guilty after mentioning Charan Ranganath—because he's the one who *taught* me about MRI physics—but I have a word limit to work with here, so I'm going to have to tap out without discussing spin echoes and *k*-space and whatnot. If you are curious, search "How does an MRI work" and "NIH" (National Institutes of Health), and you'll find a nice article with a video!

regions that are working hardest. So, as long as we can get someone to keep their head *completely* still while lying in a narrow tube and listening through headphones or viewing a computer screen through a mirror placed directly in front of their eyes, we can *watch their brains work*. I've been doing this myself for sixteen years and I still get the goose bumps when I think about it. It's. So. Cool!

In Cohen's study, people lying in the MRI scanner did a task that looked kind of like an experimentally controlled version of *The Brain Wants What It Wants*. On each trial, participants had to decide whether they wanted to choose what was behind the "safe" door—which would pay them $1.25 80 percent of the time and nothing the rest of the time, or they could make a "riskier" choice, choosing the door that *could* pay $2.50 but would only do so 40 percent of the time. Unlike the real-world version of these decisions, people in the study were *told* exactly what their odds were. However, as those of you who like to crunch numbers probably already figured out, either door would be expected to pay the same amount in the long run. So how do you think you would choose?

It turns out that neither introverts nor extraverts *like* to open a door that has nothing behind it more often than not, so both groups picked the safe bet more often than the risky bet. But even the safe decision contained some level of uncertainty—and this was what the researchers were really interested in. Just like in the real world, know-ing the *likelihood* of something happening is far from a *guarantee* that it will or won't happen.[21] What Cohen and his collaborators were re-ally interested in was the *momentary* changes in brain activation that followed the big reveal. Did I win money or not? Their results, which

---

21. I don't know about you, but I still feel frustrated when the weather forecast says that there's a 10 percent chance of rain and it rains. This is so silly of me, not only because I live in Seattle but also because I understand that statistically speaking, it *should* rain once every ten times that the forecast says there's a 10 percent chance of rain.

were replicated in two separate groups of participants, showed that the more extraverted a person rated themselves to be, the *bigger* their brain responses were to rewarding attempts compared to non-rewarding ones. Conversely, the more introverted a person rated themselves to be, the *smaller* the difference between winning and losing was in their brain.

Though this experiment did not measure neurochemistry directly, it did provide indirect evidence that dopamine communication works differently in introverts and extraverts. For one thing, the parts of the brain that were consuming more oxygen after rewards in extraverts are known to primarily rely on dopamine for communication. Among them was a smoking gun—the nucleus accumbens. This part of the brain is so tightly coupled with dopamine reward processing that it is known colloquially as the brain's "pleasure center."

The second clue was closer to a trail of breadcrumbs than to a smoking gun. In their second experiment, Cohen and the rest of the research team ran genetic assays on their participants to look for a particular version of a gene (an allele) that is known to affect the number of a particular type of dopamine receptors a person has.[22] And when they compared the brain responses of people with different versions of the allele, they found results that looked *a lot* like the differences between extraverts and introverts. One group had bigger responses to uncertain rewards than the other. Of course, this would have been a slam dunk if the people with different versions of the allele also varied significantly in their levels of self-rated extraversion. But unfortunately, Cohen's study didn't have enough participants (16 total, with 9 in one allele group and 7 in the other) to show a significant effect in this direction.

---

22. If you'd like to know more details, the Taq1A gene site is related to the expression of one type of dopamine receptor, the D2 receptor, which *inhibits* neurons in the dopamine system. We'll get into more details about why this matters in the "Navigate" chapter.

However, five years later, Luke Smillie and collaborators provided the data that closed this loop. Their experiment, which related the same genetic variant to extraversion in a much bigger group (224 participants), found that 93 participants had the version of the allele that was associated with bigger brain responses to reward in Cohen's study and were also significantly more extraverted than the 131 participants who didn't have that version!

Since then, Smillie and his colleagues have been busy collecting more evidence linking extraversion to dopamine through both genetics and brain responses. Their studies use another popular tool for studying brain functioning—one with complementary strengths to MRI. The method is called electroencephalography, commonly abbreviated to EEG.[23] In essence, EEG involves placing sensors on the scalp that are sensitive enough to measure the synchronized communication of large groups of neurons in the brain. This works because when a neuron accumulates enough evidence to send its chemical package out to whoever is listening, the process briefly changes the electrical polarity inside and outside of the cell. What's *wild* is that if enough neurons are firing at the same time, you can measure changes in electrical activity outside of a person's head! Though it's difficult to know exactly where in the brain these signals are coming from,[24] you can get very precise information about how activity in the brain changes on a millisecond-by-millisecond basis.

Because EEG technologies have been around much longer than MRI, neuroscientists have accumulated a lot of evidence about how electrical activity changes in the brain when someone opens a

---

23. This is the tool that relies on the swim-cap-style hats that got me into this business in the first place.

24. Difficult but not impossible. There are many sophisticated algorithms available that use changes in electrical activity all over the head and take anatomical constraints into account to generate models of where the "source" or "sources" of electrical activity observed over a particular region of the scalp might be.

metaphorical door to find an unexpected reward or lack thereof. To be a bit more specific, a large negative shift in electrical polarity—appropriately named feedback-related negativity—reliably follows the news about how something turned out. And EEG is much more affordable than MRI, which makes it is more feasible to have the large number of participants desired for a well-powered individual differences study.

In a 2019 study, Smillie and collaborators collected data from 100 participants, using EEG to measure individual differences in brain responses when people received news about unexpected rewards versus unexpected non-rewards. Each participant in the study also completed three different personality measures, including the Mini-Markers test that I modified for the assessment in this chapter. Consistent with the work of Mike Cohen and team, the researchers found that increased neural response to unexpected rewards was *uniquely* related to extraversion in three different personality measures but did not relate to any other personality characteristics.

Now, let's tie this all together with what we've learned about dopamine reward networks to figure out how *you* work. The results from these studies suggest that the more extraverted you are, the more strongly your brain responds when it finds an unexpected reward. This is like saying that extraverts get extra *pleasure points* every time they're positively surprised by life. This provides a chemical explanation for why extraverts also tend to rate themselves as happier and more optimistic than introverts do. And if happening upon good things feels better to extraverts than it does to introverts, is it any *wonder* that extraverts go around seeking external stimulation? As you might recall, larger dopamine responses are also associated both with stronger learning and stronger motivation to get rewarded again.

But what about the *costs* of this design feature? How could there possibly be a downside to being *extra* happy? The answer to these questions lies in the processes that are needed to override the temptation of things that make you feel good, especially when those things aren't good for you.

The strength of the "siren call" of dopamine was first demonstrated in the 1950s, when scientists put an electrode directly into the part of a rat's brain that releases dopamine when good things happen. When the rat pressed a particular lever, a small jolt of electricity was delivered to his brain, which released dopamine like crazy. As you might suspect, based on what we now know about dopamine, the rat very quickly learned to press the lever and was strongly motivated to do so. In fact, according to a *Scientific American* article written about these studies, the rats would press that magic pleasure lever as much as five thousand times per hour, sometimes for twenty-four hours without rest!

Through a series of subsequent experiments, the researchers measured just how strong the pull of dopamine was on these rats. When given the choice to pick food or the pleasure lever, the rats almost always chose the latter, even after they had gone several days without eating. And this brings us face-to-face with one of the major costs of having too big a dopamine response to good things. In the game of *The Brain Wants What It Wants,* pleasure trumps all.

But there are (too) many times in life when we need to resist the temptation to engage in something that will make us *feel good*, either because it is not good for us or because there's something even better waiting on the horizon for those who wait. This is simply *harder* for people whose dopamine responses are larger. Much like it's easier for me to say no to iced tea than to ice cream, it's easier for a more introverted person, with a smaller dopamine response, to say no to both. In fact, the same genetic variant that Smillie found to be more prevalent in extraverts has also been associated with obesity. On the other hand, not having enough dopamine has been related to anhedonia, or the inability to feel pleasure that is often associated with depressive episodes.

It's important to note, however, that no matter where you fall along this axis, dopamine is just one ingredient in a mixture of hundreds. And like many other design features of your brain, its effect on your behavior depends on a bunch of other features. In the next section, we'll discuss serotonin—a related neurotransmitter that interacts with dopamine in interesting ways.

Serotonin and satiety: Finding the balance between too much and not enough

Although dopamine reigns supreme in the court of pleasure, the extent to which having a lot of dopamine will make you *happier* depends critically on another key ingredient—*serotonin*. Without it, people high in dopamine may find themselves feeling like Lin-Manuel Miranda's vision of Alexander Hamilton—always driven to get more, and *never satisfied*. This is because serotonin is the neurotransmitter that provides the satiety signal that forms the critical "yin" to dopamine's "yang." Under many circumstances, dopamine and serotonin are known to ebb and flow in opposition. Dopamine levels rise in anticipation of something rewarding hiding behind one of life's doors, urging you forward toward good things. But then, once you've *gotten* the object of your desire, serotonin sends the satisfaction signal that is both literally and figuratively associated with feeling "full." In fact, 90 percent of the serotonin in your body is created in your digestive tract. And when the neurons in the brain use serotonin to communicate, they can have an inhibitory effect on dopamine, which decreases your wanting. Without serotonin's satiety signal, we would all be more prone to behave like goldfish, who—given the opportunity—will *actually* eat themselves to death![25]

So what happens in humans who don't have strong serotonin communication systems on board? The answer seems to depend on what's going on with dopamine. For example, impulsivity, or the tendency to act on urges without thinking about them, has been shown to increase when serotonin goes down and dopamine goes up. It's worth noting here that serotonin's role in helping people (and animals) stop and think *before* acting suggests that it can do more than just tell you when you've had enough. Much like dopamine becomes involved in predicting rewards, many neuroscientists have argued that serotonin is involved in learning to *avoid* aversive things.

But dopamine and serotonin are chemically related to each other, so sometimes the conditions that drive serotonin levels down can also take dopamine with them. High levels of monoamine oxidase (MAO)—an enzyme that can break down either dopamine or serotonin and render them useless for communication—is one example of such a condition. People who have higher-than-average levels of MAO in their neural cocktail are likely to have lower dopamine and lower serotonin levels. When both neurotransmitters are low, a person might feel the combination of being unmotivated *and* unsatisfied—kind of like feeling hungry but, at the same time, not interested in going to find food. The first drugs used to treat patients with depression worked by blocking MAOs and hence increasing the amount of both serotonin and dopamine communication in the brain.

As you might suspect, figuring out how much serotonin is in *your* mix is complicated because of the ways it interacts with dopamine. We can get some clues from your score on Dimension 2 of the

---

25. Unless you inject serotonin into their brains! In the process of verifying that goldfish *will* eat themselves to death, I came across a study in which researchers injected serotonin into either the brains or the guts of goldfish, and found that the former reduced their appetites, but the latter didn't.

characterizations assessment, a personality factor associated with anxiety or neuroticism on the negative end of the scale, and emotional stability on the positive end. Although a considerable amount of research has linked serotonin communication circuits to Dimension 2, the results are *not* as straightforward as those linking dopamine to extraversion. On the one hand, modern prescription drugs that target serotonin levels,[26] like Celexa, Lexapro, Prozac, Paxil, and Zoloft, are frequently prescribed to people experiencing *either* anxiety or depression. These drugs, called selective serotonin reuptake inhibitors (or SSRIs), work by blocking the sending neuron's ability to recycle serotonin, allowing unreceived messages to echo around longer, which increases the chances that they'll find an appropriate ear to connect with.

This might lead one to suspect that people who score more negatively on Personality Dimension 2 have lower levels of serotonin communication. However, even if we stop at this first clue, you might already notice some of the confusing things about the relation between serotonin levels and the way you feel. How can the same drug be used to treat depression and anxiety levels given the different experiences that seem to be associated with the two?

But the truth is that these drugs don't work for everyone. Depending on the type and severity of the symptoms being treated, some studies estimate that as many as 1 in 3 people don't improve following treatment with SSRIs. Hopefully, given what you've read so far, this won't come as a complete surprise to you. Mental health is exactly as complicated as the brains that give rise to it, and the symptoms of depression and anxiety are likewise as multifaceted as healthy behaviors can be. Much like the conditions that lead to

---

26. It might be more appropriate to say that these drugs have a larger direct effect on serotonin than on dopamine, because the many different levels at which dopamine and serotonin interact make it impossible to influence one without having some type of effect on the other.

ADHD might be more or less problematic depending on the larger contexts they occur in, changing the levels of serotonin communication in the brain may have different effects on how you think, feel, and behave, depending on what else is going on inside and outside of your brain. Keep this in mind in the sections that follow as we discuss research that has tried to associate differences in serotonin levels with characteristic ways of thinking, feeling, and behaving.

Given that one of the primary functions of serotonin communication is to signal to dopamine circuits that you are satiated, and that increasing serotonin through SSRIs seems to stabilize dysfunctional behaviors in many people, it's very tempting to assume that more serotonin in your neural cocktail is better. In fact, because low serotonin levels are so commonly associated with depression, many people have called it the "happy drug"! But this is *not* quite right. At least within the "typical" (or not dysfunctional) range, quite a few studies have suggested that very high serotonin levels correspond to the lower (more anxious) scores on Personality Dimension 2.

For example, a number of genetic studies have compared personality traits related to anxiety or neuroticism to variants of a gene that influences natural serotonin reuptake. People with the long allele version of this gene create 1.7 times as many serotonin reuptake gates (serotonin transporters) as those who have the short version. Because these gates work to recycle serotonin, reabsorbing the unsent messages into the sending neuron for future use, people with the short version of the allele are likely to have more serotonin echoing around in their neural cocktails. If all other things were equal, you might expect people with this version of the allele to have patterns of thinking, feeling, and behaving that resemble those being treated with SSRIs. However, one study that compared personality traits in 505 individuals with different versions of this gene found that people with the short allele rated themselves higher on anxiety-related personality characteristics, on average, than did those with the long allele![27] Despite the large number of participants in this experiment, however, the results have not been consistently replicated in other studies looking for a relation between serotonin reuptake genes and anxiety.

So what gives?

One explanation for these inconsistencies is that the characteristics of Dimension 2, or similar anxiety-related personality dimensions, have not been fully sorted out. Maybe jealousy and anxiety aren't driven by the same brain factor after all? One meta-analysis of twenty-six studies linking serotonin reuptake genes to personality constructs related to neuroticism found that results varied depending on the specific personality measure each study used. This suggests that the extent to which serotonin levels drive personality factors depends on the questions used to assess traits related to anxiety or neuroticism. Another possible explanation of the inconsistencies observed is that the relation between serotonin reuptake genes and anxiety-related personality characteristics differs depending on dopamine levels, which are typically ignored in these studies. As discussed at the beginning of this section, it is quite likely that people who have high dopamine levels experience changes in serotonin levels differently from those who are low in dopamine. A third possibility is that the relation between anxiety and serotonin depends on the external environment, or the amount of stressful or "to-be-avoided" stimuli a person experiences. In the next section, we'll explore this third possibility a bit more by summarizing the research on how different brains respond to stress.[28]

---

27. It's worth noting that although the effect of genetic variation on anxiety-related personality characteristics was significant in this study, it explained a rather small percentage of the variance—3 to 4 percent of the total variance in personality characteristics and only 7 to 9 percent of *inherited* variance were explained by this serotonin gene alone.

28. If I had limitless page space and you had limitless attention, I would try to get into the fact that serotonin's effects on you also depend on which of the fifteen

What happens when you add stress to the mix?

The first thing I'd like to point out is that stress is not necessarily a *bad* thing. Though most of us associate the word with an undesirable mental state, from a neuroscientific perspective, *stress* is a natural response that brains and bodies mount in the face of a number of different environmental demands. Whether you are experiencing a conflict with another person, a physical stressor like cold or hunger, or simply find yourself in a new or unexpected situation, your brain's stress response is intended to prepare your mind and body to react. But as I'm sure you can guess based on what you've read so far, as well as on your lived experiences, not all brains respond to stress in the same way. For example, many people view *depression* as the consequence of a brain that is not responding to environmental stressors in either a typical or functional way.[29]

Though healthy brains can respond to stress in a number of ways, all *typical* brain responses to stress involve changes in neurochemistry. The first lines of response tend to involve epinephrine and norepinephrine (also called adrenaline and noradrenaline), which prepare you for fight-or-flight responses. When these ingredients are released into the brain and bloodstream, they create a cascade of events in the body—from increased heart rate and blood sugar to decreased muscle tone in the lungs to improve breathing. When a "near miss" or startle of some kind leaves you feeling shaky and light-headed, you are feeling the effects of epinephrine and norepinephrine.

Unfortunately, many of the stressors we experience today tend to

---

different types of serotonin neuroreceptors it comes into contact with and where in the brain and body they are—but that would be a whole book in and of itself.

29. This is one of the plausible reasons why both depression and anxiety are improved when serotonin levels are adjusted in the brain—both may be inappropriate stress responses.

last much longer than the ones our brains evolved to respond to with fight-or-flight actions. And when they do, your brain responds with the "marathon" stress-response chemical—*cortisol*. From an evolutionary standpoint, cortisol's mechanisms in the body are intended to help you use your energy reserves conservatively in the *rare* occasions when prolonged stress is unavoidable. To do so, cortisol slows down your metabolism, blocking insulin's signal for your body to use the sugar in your blood as energy. But the truth is, unless you are Forrest Gump, the sustained stressors we experience today—like trying to establish financial security for our families or to survive a global pandemic—aren't readily solved by running away.[30] And because neither our brains nor our bodies were designed to withstand *chronic* stress, there are a number of negative health consequences associated with prolonged increases in cortisol levels.

So how might individual differences in serotonin levels interact with your body's neurochemical stress response? One clever experiment conducted by Baldwin Way and Shelley Taylor suggests that genes related to serotonin reuptake can influence the way people's brains respond to stress. In their experiment, genetic information was collected from 182 healthy young adults to determine whether they had the short (low serotonin reuptake) or long (high serotonin reuptake) version of the serotonin transporter allele. Afterward, all participants were randomly assigned to either a low- or high-stress experimental condition. In both conditions, participants were given five minutes to prepare a speech about why they would be a good candidate for a fictional job. The difference between the high- and low-stress conditions was that people in the lower-stress version recorded their speeches alone in a room, while those in the higher-stress condition had to deliver them live in front of an audience that

---

30. Regular exercise *can* help, though! Stay tuned.

was judging their performance.[31] To record their participants' stress responses, the researchers also measured the amount of cortisol excreted in their saliva at the beginning of the experiment, and then again twenty, forty, and seventy-five minutes later, with the last window being well after the speech had been completed. As expected, cortisol levels didn't change as much, *on average*, in the low-stress condition as they did in the high-stress condition. But critically for the discussion at hand, people with the short alleles—those whose serotonin signals stay in their synapses longer due to less-effective recycling—experienced *significantly higher increases* in cortisol levels in the high-stress condition than did those with the longer alleles. In the low-stress condition, however, the two genetic groups didn't differ in their cortisol levels. Taken together, these findings suggest that the relation between individual differences in serotonin reuptake and self-reported anxiety characteristics is probably more salient when people are exposed to high levels of environmental stressors than when they are not.[32]

The moral of this story should be starting to sound familiar to you—none of your brain's design features exist in a vacuum. On the contrary, each one is engineered to prepare the appropriate response to some class of environmental triggers. In "Lopsided" we discussed the large-scale organization of the two hemispheres of your brain and how differences between them can interact with your level of experience to shape the way your brain solves complex problems. In this chapter we discussed how different neurons organize themselves

---

31. This manipulation is commonly used in stress research. Apparently, enough people find either public speaking or being judged stressful enough that the one-two punch of them combined stresses most participants out.

32. Based on a study conducted on over 21,000 Finnish twins that showed changes in self-reported neuroticism levels following significant life stressors, I'd be pretty surprised if the "average" responses on these anxiety- or neuroticism-related personality scales haven't shifted since the global pandemic began in 2020.

into functionally specialized networks by speaking specific chemical languages and how different environmental triggers can exacerbate individual differences in the way these networks communicate. In the next chapter we'll discuss one last design feature that shapes the way your neurons connect with one another to build an understanding of the world around you—one that drives how flexibly your brain can respond to the same environmental triggers when they are encountered in different circumstances. But first, let's summarize the key ideas we've covered in this chapter and what they might suggest about how your brain works.

Summary: Differences in your brain's chemical checks and balances shape your decisions and responses to environmental pressures

Before I remind you about the details of what we discussed in this chapter, please allow me to provide some big-picture context for your right hemisphere to use to figure out what it all means. First, I'd like you to keep in mind that of the *hundreds* of neurotransmitters that drive your characteristic ways of thinking, feeling, and behaving, I have only discussed three in any detail—dopamine, serotonin, and cortisol.[33] But even in this dramatically simplified problem space, you can start to see how things get complicated quickly, because the influence that each of these ingredients has on *you* depends critically not only on what environment you find yourself in but also on the functioning of other communication systems in your brain.

---

33. We'll talk about another important neurotransmitter—oxytocin—in the last chapter of the book.

And there's something else I haven't mentioned yet that further complicates this delicate balance—the fact that your brain has ways of adapting to changes in your neurochemistry that help it *maintain* its preferred chemical levels. At the beginning of the chapter, I discussed four different design features that drive your chemical communication systems—the amount of neurotransmitter available to your sending neurons; the number of receptors, or "ears," available to receive a particular type of chemical message; the amount of reuptake, or recycling, that a sending neuron is capable of; and the amount of enzymes available for breaking down that neurochemical. One thing that too few people realize is that when you alter your neurochemistry artificially—whether it be by drinking lots of caffeine or by taking a prescription drug like Prozac—your brain will often respond by changing its other design features to try to counteract the effects. For instance, to counteract the effect of drugs that *increase* the levels of dopamine in your neural cocktail, your brain might reduce the number of receptors it has for receiving dopamine messages, or increase the amount of enzymes in your brain that render dopamine useless for communication. Not only does this create resistance to the effects of drugs that alter neurochemistry over time, it can also cause a variety of withdrawal symptoms if you stop taking the drug. After accommodating to the effects of multiple cups of caffeine every day, for instance, your brain starts to recalibrate so that a certain amount of caffeine is required for "normal" functioning. As a result, people who reduce their caffeine intake dramatically often experience a number of symptoms—including headaches, difficulty concentrating, and depressed moods—which reflect the way their brains have adjusted to its chemical interventions. Thankfully, these symptoms only last a few days (two to nine days, on average), as most brains will readjust to the *new* baseline levels again once the drug is removed.

Having at least a rudimentary grasp on these dynamics is important for figuring out how differences in neurochemical communication systems shape the way you experience the world around you. In this chapter, I introduced the idea of dopamine reward circuits, and how people who describe themselves as being more extraverted are likely to have brains that produce more feel-good dopamine rewards when unexpectedly good things happen to them. This, in turn, increases the likelihood that they will repeat the actions that led them to those unexpected rewards in the first place. But too much dopamine—especially in the absence of satisfaction-inducing serotonin—can lead someone to overindulge in unhealthy or addictive behaviors. On the other hand, too little dopamine has been associated with low motivation, or with the inability to feel pleasure, which lies at the heart of many cases of depression. In the "Focus" and "Navigate" chapters, we'll discuss the role that dopamine plays in attention, motivation, and decision-making in more detail.

We also talked about how, under ideal circumstances, dopamine and serotonin work in combination like yin and yang in your brain. To prevent you from behaving like a dopamine-dependent rat that pushes a lever for twenty-four hours straight to get rewards (without stopping for food or rest), our serotonin systems *can* kick in and provide the "satiety" signal that inhibits dopamine and quenches our desire for more. Ironically, however, too much of the "I've had enough" neurochemical has been associated with higher anxiety levels, especially when people are placed in stressful events. But given that your brain may counteract the effects of any drugs you take to alter it, how might you achieve balance if your yin and yang are out of whack?

The first thing to consider is your nutrition, because some of the building blocks your brain needs to *make* neurotransmitters aren't produced internally. They have to be ingested. One example of this is *tryptophan*, an amino acid found in relatively high concentrations in poultry, eggs, fish, and milk. Tryptophan is one important precursor for building serotonin, and it can't be produced by the body. In fact, one way neuroscientists study the effects of *low* serotonin levels on behavior is to feed participants diets that are high in amino acids but don't have *any* tryptophan.[34] *Tyrosine*, the corresponding precursor to dopamine production in your brain, is found in high concentrations in cheese[35] but is also found in other dairy products, as well as in the aforementioned poultry and fish. Tyrosine is a common ingredient in many pre-workout or energy drinks, as it is *also* a precursor for building norepinephrine. But proceed with caution; research studying the effects of tyrosine on blood pressure and anxiety levels has yielded inconsistent results.[36]

Never fear (pun intended), there is also a host of healthy *activities* that can reduce stress levels *and* adjust your chemical ingredients. For example, moderate levels of aerobic exercise have been shown to *increase* both serotonin and dopamine levels in the short and long term. In their 2016 review article, Saskia Heijnen and colleagues describe the results of a number of different studies suggesting that the physical stress experienced by people who exercise (which they term "good stress") can result in very different, typically more desirable, long-term effects on the brain than prolonged psychological stress (or "bad stress"). The precise mechanisms that link

---

34. This protocol, called *acute tryptophan depletion*, has been shown to temporarily reduce the amount of serotonin communication in the brain by as much as 90 percent! But please don't try this at home just for funsies. Some of the effects of tryptophan depletion include depressed mood, exacerbated PMS, and slowed gastric emptying, just to name a few of the most exciting ones.

35. In case you needed another reason to love cheese—you're welcome.

36. These inconsistent findings probably reflect similar complications to those we've discussed in this chapter—interactions with environment and pre-tyrosine neurotransmitter levels, to name a few.

exercise to changes in neurochemistry are multifaceted and are still under investigation in human and animal models. For example, increased serotonin levels in the brain may arise, in part, because when active muscles absorb long-chain amino acids from the bloodstream, they increase the opportunity for tryptophan to cross the blood-brain barrier. Increased dopamine levels, on the other hand, have been associated with the release of endocannabinoids following exercise—the naturally occurring neurotransmitters that drugs like THC (found in cannabis) mimic in the brain.[37]

Other activities that target stress reduction have also been shown to change neurochemistry. Massage therapy, for instance, might have just the effect on your mixology you've been looking for! Not only has it been repeatedly shown to reduce levels of cortisol in your brain, sometimes by as much as 50 percent, it also increases dopamine and serotonin levels by up to 40 percent. Meditation and mindfulness practices have also been associated with reduced cortisol and increased serotonin levels. One study even found that people who practiced deep-breathing exercises reduced their cortisol levels and improved their mood. Of course, consistent with one of the central themes of this book, the extent to which any of these "interventions" will create a meaningful change in *your* brain depends on what else is going on. Baseline neurochemistry levels as well as differences in life circumstances can moderate the effects that any of these practices have on your brain.

Though I worry a little that you might be getting tired of hearing answers that boil down to "it depends," the truth is that these kinds of interdependencies in your brain lie at the heart of your brain's superpowers. Because if your brain *weren't* engineered to respond

---

37. This provides a whole new level of appreciation for what it means to have a "runner's high."

differently in the face of changes in its internal and external environ-ments, *you* would be more predictable, less interesting, and a whole lot less successful in evolutionary terms. In the next chapter, we'll discuss one last biological design feature that is critically tied to your ability to behave flexibly depending on the circumstance at hand: your neural synchronization.

CHAPTER 3

IN SYNC

The Neural Rhythms That
Coordinate Flexible Behavior

The next design feature we'll explore relates to the coordination of your brain processes. Though I'm not *strictly* talking about the "pat your head while rubbing your stomach" kind of coordination, your brain's internal synchronization mechanisms are precisely what make this seemingly basic task so difficult. To get a better handle on how this works, let's return to the idea of the game Telephone. As you probably remember, one of the critical challenges in your brain's version of the game is that there is *a lot* of background noise created when thousands of neurons in close physical proximity release their chemical packages. In "Mixology," we talked about how brains use different combinations of "chemical languages" as one way of imposing order on the massively overlapping signals between neurons. In this chapter, we'll discuss another way brains can manipulate which neurons are talking to one another—one that allows those that speak the same languages to be assembled into different teams *dynamically*, depending on the job at hand.

To put it simply, another trick your brain can use to influence which signals will be "heard" among the background noise is to coordinate the timing of sent messages. To better understand how this works, think about the differences between the sound of a party and the sound of a chorus. When you first walk into a party, the cacophony of dozens of raised voices engaged in different conversations reaches your ears like a big, indistinguishable racket. Trying to "tune in" to a single conversation against that kind of background noise is challenging at best. Now compare that to the synchronized voices of a choir, which blend together to create sounds that can be heard and understood much more easily—even over background noises coming from the audience.

Signaling in the brain works like this as well. If two messages are delivered at the same time, they are much more likely to be "heard" by, or to have an effect on, a receiving neuron, than if their packages are received out of sync. And it turns out that your brain, like many natural phenomena, is a massive rhythm generator. Rather than sending continuous signals,[1] individual neurons cycle through "whisper" and "silent" phases, and can do so at different frequencies. Receiving neurons can also be engineered to "tune in" to a particular frequency, much like selecting a radio station on your car stereo allows you to hear only the waves transmitted through the air at a particular frequency. And though all brains use frequencies that range from less than one to well over one hundred signals per second, individual brains differ in how much of their communication occurs over the slower or faster frequency ranges.

In fact, we've been measuring the makeup of this "orchestrated" neural communication in the lab for quite some time now, using the same electrode caps I practiced with on Jasmine twenty-five years

---

1. This is impossible, not only because neurons would run out of their chemical packages quickly but also because of the way action potentials work. After "firing" off their chemical message, neurons need a fraction of a second to reset themselves before firing again.

ago.[2] But unlike the experiments we've talked about previously, which measured changes in the electrical activity of the brain associated with a *specific* task, this research looks at brain activation collected in the *absence of any task*. Participants in my studies are given the straightforward (though not always easy) instruction to close their eyes and relax—without falling asleep. When they do, their minds are free to wander (like mine on bus rides). And while that happens, we record the ebb and flow of electrical activity that their undirected brains create, over a period of five to ten minutes.

Then, to figure how each individual's brain is orchestrated, we mathematically decompose the changes in electricity recorded over the scalp into different frequency bands. This is kind of like listening to your brain's choir and trying to decide how many sopranos, altos, tenors, and basses are singing in it. Though there are a few different ways of doing this with EEG data, the result is generally the same— an estimate of how much of the activity recorded over a given region is coming from neurons with synchronized communication in a particular frequency range. I've included an example of what this looks like when the data recorded from *my* wandering brain was decomposed into frequencies ranging from 2 to 40 Hz, or cycles, per second.

---

2. Fortunately, this is much easier to do with adults. And some of the "consumer-grade" equipment we use looks more like a Space Age headband than a swim cap. No gooey gel required!

CHANTEL'S NEURAL ORCHESTRATION

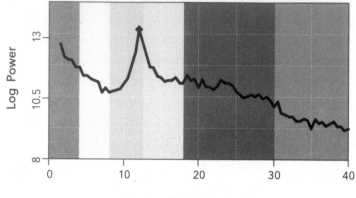

Neuron Firing Rate (in Hz)

The height of the line in this graph represents an estimate of the proportion of my brain communication that is occurring at a particular frequency. The higher the line is, the greater the proportion of communication is at that frequency. And as you can see, that proportion rises and falls as you move from the left side of the graph, which measures the lowest frequencies of communication, to the right side of the graph, which represents the highest frequencies, with a sharp peak right around 12 Hz. That peak, which is marked with a little diamond, is my brain's preferred channel to use for mind wandering. The *height* of that peak shows that there are a lot more neurons in my brain sending signals at a rate of around 12 messages per second than there are at, say, 10 or 15 messages per second. We'll talk more about what this means later in the chapter. Critically for understanding how your brain works, these measures of "task-free" brain orchestration are kind of like a neural fingerprint. They are relatively stable within a person but vary substantially between people. However, since we can't send electrode caps home to you (yet), we'll have to rely on what we've learned about how these frequency fingerprints relate to the way brains perform different kinds of computations.

But first, even though it makes me feel a little vulnerable, I need to point out that having a more synchronized brain is *not* always a good thing. A dramatic example of this is the epileptic seizures that can be triggered when one brain region's firing causes a cascading electrical storm that travels, unchecked, throughout the brain. It's actually kind of amazing that this doesn't happen more frequently than it does, because *every* neuron in your brain can eventually be connected to every other neuron by an average of six intermediate links (following the same principles of *network theory* that allow us to link every actor to Kevin Bacon by six degrees).[3] To prevent one gossipy group of neurons from creating mass hysteria in your brain, it is often just as important to *stop* the spread of activation between neurons as it is to get them *in sync*.

To illustrate why this matters in a healthy brain, here's another at-home brain experiment for you to try: First, rotate the ankle of your dominant foot so that your toes travel in a clockwise circle, and keep that circle going. Then take your hand from the same side and make a large number 6 in the sky on an imaginary chalkboard, as if you were writing the answer to the question "How many links are there, on average, between two neurons in the brain?"

What happened?

For the majority of us, the hand's trajectory will hijack the movement of the foot, reversing the direction of its circle. Now try the experiment again. This time, start by rotating your ankle counterclockwise before making the number 6 with your hand. What happened that time? Most of you will find this task somewhat easier. And I'm willing to bet that you may find your foot and hand wind up *in*

---

3. In case you never played Six Degrees of Kevin Bacon, the game goes like this: I give you the name of any actor, and you try to link them to Kevin Bacon by naming films that shared a co-star. Here's an example from Wikipedia: Elvis Presley starred in a movie (*Change of Habit*) with Edward Asner. Edward Asner was in *JFK* with Kevin Bacon. Elvis and Kevin Bacon are separated by 1 degree (Edward Asner).

sync with respect to their positions around the circle that ends the six.[4] This experiment provides an example of what happens when the "message" your brain used to coordinate your hand movements interferes with the message sent to the foot-control regions.

To prevent this kind of interference from happening *all the time*, your brain has another tool it can use to coordinate communication, particularly between groups of neurons that are not physically close to one another. To do this, it uses bundles of long neurons that work like information superhighways, creating bridges between distant regions of the brain. These *white matter* pathways get their name because they are covered in myelin, a fatty insulating layer that speeds up their signaling process and reduces the likelihood that any information will get lost along the way. In fact, signals that travel along such insulated neurons can reach speeds of over 250 miles per hour, allowing them to move from one side of the brain to the other in as little as eight milliseconds. By comparison, the non-insulated, *gray-matter* neurons that cover your brain transmit signals at speeds of 1 to 4 miles per hour. This is slower than my jog/walk pace—so it's a good thing these signals don't have far to go!

By now, I hope you know me well enough to guess that having faster brain communication isn't always better, even if it does seem appealing. But given the speed and efficiency of the white-matter superhighways that make up more than half of the adult brain, why would evolution even bother with the slower and noisier "rhythmic" ways of coordinating brain signals?

The answer, in a nutshell, relates to *flexibility*.

When the brain is *hardwired* so that signals are quickly and auto-

---

4. Note that this experiment assumes that you will begin drawing the 6 at the top and end with a counterclockwise circle. It's perfectly plausible that some of you, especially left-handers, may have learned to draw a 6 by starting at its "belly button" and drawing a clockwise circle that ends with a long tail at the top. If you do, the counterclockwise foot rotation will probably be more difficult for you!

matically transmitted from one part to the next, there is very little opportunity to *reconfigure* the flow of information. Reading provides an excellent illustration of this idea. As a person learns to read— which, for sighted individuals, requires *thousands* of hours of practice mapping squiggly lines on a page to their linguistic equivalents— their brain builds a white-matter pathway that connects the neurons responsible for visual pattern detection to those involved in retrieving the meanings of words. Remarkably, once this pathway is fully developed, fluent readers can't *prevent* themselves from reading when they see words.

The "siren call" of word reading can be demonstrated using a simple color-naming task.[5] Most toddlers with typical color vision learn to identify colors much earlier than they learn to read. So you might *imagine* that adults with typical color vision would have no problem whatsoever naming the color of ink that a word is printed in, if asked to do so in a laboratory experiment. As it turns out, this task is wickedly difficult in one particular circumstance—when the word being identified spells the *name* of a different color than the one it's printed in. For instance, if I were to ask you to name the color of ink that the word "BLUE" is printed in in this book, it would be much harder for you to say "BLACK" than if I were to ask you what color this string of X's is printed in.[6]

In the lab, we measure how quickly and accurately people can name the color of strings of letters that don't make up words, like "XXXXX," or of words like "BUMBLEBEE"[7] that are not color

---

5. This task is commonly called the Stroop Task, after the psychologist who created it almost a hundred years ago—John Ridley Stroop.

6. OK, unless book printing changes dramatically in the next year or two, this task will be pretty easy, since *all* of the words in this book are printed in black. But if you were doing the experiment in the lab, the words would be presented in different colors, which would make it a lot harder.

7. This would actually be a terrible word to use in an experiment, because it has a lot of syllables and is a low-frequency word—but it's also a fun word to say, so I chose it for my example.

words at all, compared to how quickly and accurately they can name the colors of words like "BLUE" that are printed in a different color ink. The most straightforward way to do this task would be to ignore the words altogether, as their meaning is irrelevant to the task at hand. But most fluent readers simply can't do this. Although there are interesting individual differences we'll return to in a bit, almost everyone is slower and less accurate to name the color of an inconsistent color word than they are to name the color of a non-color word. Once that information superhighway has been set up, the data coming in from your visual system gets delivered to your word-meaning region, whether you like it or not.

And this is where our neural rhythms come into play. Because one of the most remarkable characteristics of our human brains is how flexible they are. Under many circumstances, we *can* respond in different ways to the same inputs, depending on what our goal is.

Take reading this paragraph, which I believe to be one of the most important ones in the book, as an example. Hopefully, now that I've told you that this is an important paragraph, you're reading a bit differently. Maybe you are focusing your attention more strongly on the message conveyed in my words and what they might reflect about how your brain works. But what if—instead—I had started by asking you to *proofread* this paragraph? The focus of your attention would be different: a poorly placed, comma, or an incorrectly used semicolon;[8] would be more likely to capture your attention than the meaning of the ideas that are being communicated would. (OK, now stop proofreading. I want you here with me!) I could even give you an instruction about how to pronounce the words you read on the page differently: Every time you encounter the letters *th*, pronounce them like *d* in your head. And most of you will be able to do this without a

_____

8. Yes, those were intentional. Any other typos or grammatical errors in this book—not so much.

problem, despite the fact that you've never had a reason to do it before. Now your "inner" reading voice has a lovely accent!

What I've illustrated here is the remarkable capability of your brain to reprogram itself, on the fly, based on changing goals or instructions. That is to say—after the highly automatic process of word reading has been accomplished, your brain can use the *message* conveyed in the words to reconfigure its order of operations and do *different* things. This dynamic reconfiguration of neurons into different teams, depending on the instruction or goal at hand, is simply not possible when pathways have been hardwired with white matter. It requires the careful orchestration of your brain's choir over different neural rhythms. In the following section, we'll discuss what we've learned about the role of various neural rhythms in this type of flexible coordination, along with the costs and benefits of your brain's preferences for thinking fast and slow.[9]

The costs and benefits of different speeds of neural synchronization

Before explaining what we've learned about how people with different methods of neural orchestration work, we'll need to discuss the different types of computations that fast and slow rhythms are best suited for. The first thing to note is that it tends to be *easier* to keep groups of neurons synchronized at slower frequencies than it is at faster ones. Returning to the metaphor of a choir, you might imagine that the slower a song is, the easier it is to keep the group of voices close *enough* together in time that they still *sound* like a unified whole. What would happen if you tried to assemble a choir of the fastest

---

9. Not to be confused with the famous book by Nobel Prize–winning psychologist Daniel Kahneman—but as you'll read, there *is* a relationship between the two ideas.

rappers in the world, many of whom can produce speech at rates of over twelve syllables per second? Even if they were a *fraction* of a second off, the group would quickly blend into an unintelligible mess. For related reasons, low-frequency brain waves tend to be involved when larger groups of neurons are coordinated. For instance, when you are in the deepest phase of sleep, most of your brain oscillates together in synchrony over the slowest frequency range—less than four signals per second.

Another thing to note is that low-frequency waves are capable of traveling farther and are less perturbed when they bounce off things in the environment than higher-frequency waves are. This is why the rumble of an elephant call, which occurs at around the same preferred frequency range of my brain (about 12 Hz), can be heard by other elephants more than two miles away! In comparison, the chirp of a bird, which can range from 1,000 to 8,000 Hz, is much less likely to be detectable over long distances.

And what do you think happens when the rumble of a large group of neurons communicating at slower, elephant-like frequencies in the brain clashes with the chirps of a small flock of birds? Though colliding signals can create complex and interesting dynamics, the net result is typically that the lower-frequency brain waves can have a large effect on communication in the higher-frequency ranges, while the reverse tends not to be true.

So what's the point of a high-frequency "bird chirp" communication system if it gets bowled over by lower-frequency signals? For one thing, neurons communicating at faster frequencies can update their representation of what's going on around them more rapidly. As you might remember from our discussion in the "Lopsided" chapter, many everyday tasks, like language comprehension, rely on the ability to detect changes in the environment with millisecond-level precision. Can you imagine how difficult it would be to survive if your brain were communicating about what's going on in the world only a couple of times per second? In fact, our *sensory* brain networks tend to talk among themselves at the fastest frequencies. And the faster your brain synchronizes its communication in these networks, the closer it will be to having "real-time" representations of what is going on around you.

So what might the rumbling, slower frequencies be doing? As you may have guessed, these wavelengths are ideal for *assembling* different parts of the brain into teams. In fact, the research of Earl Miller and his collaborators has shown that there are important neurons in the frontal lobe that can orchestrate the higher-frequency neural processes in the rest of your brain dynamically, based on the goal you're trying to accomplish. Such coordination *imposes* order on the cacophony of overlapping messages in the brain, like conducting a choir made up of 86 billion voices.

Obviously, having a flexible, coordinated brain is *good*—but here's the catch. First, as the foot-coordination experiment demonstrated, these lower-frequency, goal-directed signals are more likely to interfere with one another than are the faster signals coming from the local processing centers. And this, my friend, is one reason why we *suck* at multitasking.

Second, when your lower-frequency, or "goal-oriented," neural oscillations get involved in orchestrating your higher-frequency ones, they can quite literally prevent you from noticing what is happening in the world around you.[10] In the laboratory, we can measure this trade-off between being directed by your inner-world goals and your ability to detect what's happening in the world around you using a measure called the *attentional blink*. In short, individuals in these experiments are given a goal—to detect a certain type of target amidst a stream of rapidly presented visual stimuli. For example,

---

10. The different mechanisms brains use for noticing things will be the focus of the next chapter!

they might be told to remember the numbers presented in a series of stimuli that are mostly letters, or the colors of circles presented in a series of shapes that are mostly squares. The stimuli are then flashed one at a time, in quick succession. At the end of a trial, which might be made up of ten to fifteen items, people are asked to report which targets they saw. What I find fascinating about this is how the data show that there is a window of time after the first target is presented in which people don't *notice* the second target. In fact, this so-called attentional blink can last as long as half a second!

So here's the conundrum—to process the world in something close to "real time," we need brain networks to communicate over high-frequency wavelengths. But if we want to send that information to different parts of our brains, we've either got to create hardwired, inflexible circuits or rely on lower-frequency communication systems that are more prone to getting their wires crossed. Every brain uses some combination of both, but research in our lab and others has shown that the extent to which your neural communication systems are primarily orchestrated over lower- or higher-frequency wavelengths has wide-reaching implications for how different people process information. In the next section, I'll provide a few tests to help you figure out how your brain is orchestrated.

Assessing your tempo

I'm not going to lie—the first test I'm going to give you is hard! And I know that no matter what I say, those of you who do well on this test are going to feel like rock stars, and those who struggle are going to feel like you failed. But bear with me. I am pushing your brain to the limit for a reason. And I promise, as the costs-and-benefits section illustrated, there are positive consequences of being *either* good or bad at this test.

The test measures one of the biggest bottlenecks in human information processing—working memory capacity. *Working memory* is a term that describes a very privileged state of conscious awareness in which the contents of your thoughts can be used to orchestrate your mental and neural processes. People vary in how much information they can maintain in this state—a value called working memory "capacity." You might think of this as the music the conductor of your neural choir uses to orchestrate your brain's communication.

This test will measure how much information you can get into, and manipulate inside, your working memory. Because this test is designed to measure the limits of your information processing, most people will *not* get all of the items correct. Also, though you can do this yourself in the book, it'll be more accurate if you can get someone else to read the items to you, or do it on my website. Finally, because this test is really sensitive to your current brain state, you should take it when you're feeling focused and well rested!

The goal of this test is to remember as many numbers or letters as possible from a list presented to you one item at a time. The catch is that you need to recall them in the *reverse* order that they were originally presented. First, grab a pen or pencil and a piece of lined paper, and write the numbers 1 through 14 in a column to help you keep track of which line you're on. Next, if you've got an "accomplice" to help test you, have them read the letters or numbers in each line aloud at a pace of about one item per second (think "one Mississippi" between items), and then say "GO" when they've reached the end of the line. If you're testing yourself, read each letter or number only once, then flip your book over when you get to the word "GO." Hearing or reading the word "GO" will be your cue to recall the items in reverse order and write them down on your paper. For instance, if you hear or read the line "C K R G go" you would write "G R K C." If you're working with a partner, be sure to let them know when you're ready for the next question.

Note that the lines start rather short and increase gradually in length. You might get the feeling at a certain point that you have reached your capacity. If you have completely guessed on two or three lines in a row, feel free to stop the test there and end the torture. You don't get points for partial credit, so if you're just repeating back the last two or three letters over and over, you've also likely hit your capacity. A few more rules before you start: (1) don't start writing before you hear or read the word "GO"; (2) do not write the letters in the order you've heard them and then reverse them—the reversing bit needs to be done in your head; and (3) don't check your answers until you're done. I don't want you to psych yourself out. Ready to get started?

WORKING MEMORY TEST

1. 5 8 2 *GO*
2. L D R *GO*
3. 3 9 4 1 *GO*
4. D X K Q *GO*
5. 7 4 2 9 5 *GO*
6. Y M R K V *GO*
7. 4 1 8 5 9 3 *GO*
8. H D N B R T *GO*
9. 8 5 4 2 1 6 3 *GO*
10. G L Z K V I C *GO*
11. 9 4 2 1 5 8 3 7 *GO*
12. F B V K W L P S *GO*
13. 2 5 8 4 1 7 9 3 6 *GO*
14. C X S V R N D H P *GO*

When you've completed the test, we can figure out what your working memory capacity is. First, each line only counts if you got all of the items correct. As I mentioned, there's no partial credit on this test! Next, if you didn't get any of the lines completely correct, your working memory capacity is 2. Otherwise, find the longest line you got correct for *both numbers and letters* to calculate your score. If you got only one of the two lines of a certain length correct, give yourself a .5 increase. For instance, if you got both lines that were three items long correct, but only one of the four-item lists, your working memory capacity is 3.5. As a sanity check, your final score on this test should fall somewhere in the range of 2 to 9.

OK, now that I've exhausted your brain with all of this letter and number juggling, let's try something a bit more fun. To do it, you'll need a pencil and a few sheets of paper, and a timer of some sort. The task is loosely based on the figural component of the Torrance Tests of Creative Thinking. Your goal is pretty straightforward—draw as many objects as you can in five minutes that contain the shape on the next page (don't peek!). Try to think outside the box when you do this, because you get bonus points for being original. But also try to produce as many drawings as possible. The goal is both to produce as many different things as you can and to produce as many *original* things as you can, in the five-minute time period. Once you've got your equipment and the timer set, hit Start and turn the page.

CREATIVITY TEST

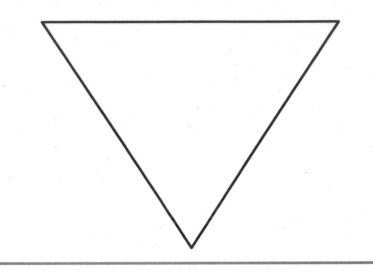

When your time is up, count how many different pictures you created in the five-minute time window. In the lab, this test would also be scored for how original each item is, by comparing your drawings to the drawings of other people in the experiment. I'm willing to bet, for example, that many people will draw something like a tent, an ice-cream cone, and—if you thought to flip the shape (which was not forbidden)—a house or a person wearing a party hat. But fewer people would draw things like a fox, a stegosaurus, a volcano, or a martini glass. For now, just count the number of things you drew, and make your best guess about how unusual your responses were. We'll talk about what this tells us about you in a bit.

But first let's try one more test. This one kind of walks the line between the first super-hard test and that second one, which was more fun and creative. Some of the items in this word-puzzle test are certainly hard, like the working memory test, but—in my opinion— they're much more fun to solve! The items on this test were borrowed from the Compound Remote Associates Test created by Edward Bowden and Mark Jung-Beeman. Each question will contain three words. Your goal is to find a fourth word that links the three other words together by creating familiar phrases or compound words. For instance, if I were to give you the list "cottage, swiss, and cake," can you come up with a fourth word that forms a meaningful combination with all of the other three? The answer in this example would be "cheese," because "cottage cheese," "swiss cheese," and "cheesecake" are all things.

You'll also need a pen or pencil, a piece of paper, and a timer for this test. There are ten items, so you can number the paper from 1 to 10 to help you keep track of where you're at. I'd like you to spend no more than thirty seconds trying to find the answer to each problem. But there's also a catch. I'd like you to take a note about how the answers come into your mind. If the fourth word seems to "pop" into your head in an "Aha!" type of experience, make a little + mark next to your answer. If, instead, you have to search for it more systematically, for instance by thinking of all the words you know that go with "cottage" and trying them out on the other words, put a check mark next to your answer. These problems get harder as you move along, so don't worry if you don't figure them all out. I've included a range of difficulty intentionally.

COMPOUND REMOTE ASSOCIATES TEST

1. sandwich/house/golf
2. rocking/wheel/high
3. worm/shelf/end
4. basket/eight/snow
5. hammer/gear/hunter
6. man/glue/star
7. baby/spring/cap
8. note/chain/master
9. over/plant/horse
10. service/reading/stick

When you're done, turn the page to see how many you got correct!

ANSWERS TO COMPOUND REMOTE
ASSOCIATES PROBLEMS

1. club
2. chair
3. book
4. ball
5. head
6. super
7. shower
8. key
9. power
10. lip

After checking your answers, please look *only* at your correct answers and calculate your *insight ratio* by dividing the number of correct items that were associated with an "Aha!" insight moment by those that were solved by working through the space systematically. Since you can't divide a number by zero, replace the denominator with 1 if you didn't solve *any* of the problems using a systematic method.

How typical are you?

Now that I've worn out all the corners of your mind and brain, I hope you're curious about what your performance on these tests has to do with how your brain is synchronized. In fact, each of these tests has been linked, in one way or another, to individual differences in your brain's dominant frequency of communication, which is appropriately named *alpha*.[11] In my brain, the preference is shown by the sharp peak at 12 Hz. As a first step, let's figure out what these tests might tell us about your brain's preferred tempo.

First, we're going to use your score on the working memory test to get an estimate of how quickly your brain likes to oscillate—at least when you're in a relaxed, mind-wandering state. Or, as J. K. Simmons says in the movie *Whiplash*, "Were you rushing or were you dragging?" Based on a large study by Richard Clark and colleagues that measured alpha frequency and working memory capacity in 550 people from three different countries, the more items you can hold in your working memory, the faster your alpha frequency is likely to

---

11. In truth, the alpha frequency range got its name simply because it was the *first* neural oscillation to be discovered. This isn't surprising, given how prevalent it is in the brain. When a person is in a relaxed state, you can often see alpha waves with the naked eye in their EEG recordings—no fancy mathematical decomposition required.

be. According to their data, the average score on this working memory test is right around 5. To be a bit more specific, for those of you between the ages of eleven and thirty, about 68 percent of you will have scored between 3.5 and 7 and will have corresponding peak alpha frequencies ranging from 8.5 to 11 Hz. Though the likelihood of scoring highly on this test decreases with age,[12] results from this study showed that the relation between alpha frequency and working memory capacity is consistent across the age span. Specifically, the authors claim that each increase of 1 Hz in a person's alpha frequency corresponds to the ability to hold .2 more items in mind, on average. So for the remaining 32 percent of you with working memory spans that are in a higher (7 to 9 items) or lower (3 items or fewer) range, you are also likely to have preferred, or "peak," alpha frequencies that are greater than 11 Hz or less than 8.5 Hz, respectively. And as you've learned by now, the closer you move toward the extremes, the less typical your pattern of brain orchestration is likely to be.

Put a pin in the creativity test for now; we'll loop back to it.

Let's talk for a minute about your performance on the Compound Remote Associates Test. The general difficulty of these problems increases as you move from the top of the list, which includes items that were solved by more than 80 percent of college students within the thirty-second time window, to the bottom of the list, which were solved by 10 percent or fewer of them. At least one study has shown that people's ability to solve these problems is correlated with their working memory capacity and fluid reasoning abilities. So those of you who found the answers to most of these problems also probably scored above 5 on the working memory test.

But what I'm *really* interested in here is what percentage of the

---

12. On average, the age group of thirty-one to fifty, which I proudly occupy, remember about one fewer item, on average, than the younger group, while the fifty-one-and-ups remember an average 1.5 items fewer than the thirty-and-unders.

time you landed on the answers you *did* get correctly by an "Aha!" insight moment versus through a deliberate search process. According to a recent study Brian Erickson and collaborators conducted on fifty-one young, right-handed adults, the extent to which people arrive at the answers to problems like these via insight versus deliberate search processes seems to be a rather stable characteristic. To assess this, they gave their participants different types of problems, including compound remote associates puzzles like the ones I gave you, and anagrams—scrambled words or phrases like "RANIB" that need to be reshuffled to spell a real word—approximately two weeks apart. Much like I did previously, they also asked their participants to report whether they found each answer through a deliberate search or a moment of insight. Interestingly, they found a significant positive correlation between the two tasks. Specifically, the percentage of time a person relied on deliberate problem solving for one type of task explained almost 30 percent of the variance in how frequently they would rely on the same strategy to solve a totally different problem, weeks later.

Critically for understanding how your brain is orchestrated, the different problem-solving styles were associated with distinct patterns of neural synchronization extracted from the task-free brain activation of these participants. Specifically, people who tend to rely on insight had greater proportions of their communication happening over low-frequency wavelengths from 4Hz to 14Hz, particularly over the left-hemisphere language regions.[13]

Taken together, these two tests provide complementary information about how *fast* your dominant neural rhythm is, and how *strong* your synchronization is over low- versus high-frequency channels of communication. Note that the two tests we've discussed so far

---

13. As a reminder, these were all right-handers, so I'm assuming that most of them had left-dominant language functions.

provide different types of information. Returning to the graph of my brain waves, you can imagine that a person with a high working memory capacity might have a very fast alpha peak—which would correspond to how far to the right of the graph that peak labeled with a black diamond falls—and still have either high or low amounts of communication happening in the low-frequency channels, which would correspond to the height of the peak as well as the height of the line to the left of the peak. In the next section, we'll talk a bit more about the broader implications of having brains with these different communication profiles.

How do you tune in?

To get a better understanding of how your neural orchestration influences the way you think, feel, and behave, I'll need to explain one last thing that makes your brain's version of Telephone a lot more complicated than the children's game. Remember that in the children's game, one message travels around the circle, beginning and ending with the same "sending" child. But in your brain's version of the game, two messages can be traveling around the circle in different directions! One type of message comes from the neurons that make contact with the outside world through your senses and communicate at higher-frequency wavelengths. We call these signals "bottom-up" processes, because their goal is to guide your understanding of the world and your decisions about it from the ground up, based on their best understanding of what is happening around you in real time. The other type of message originates in your frontal lobe "control center,"[14] where lower-frequency channels can be used to

---

14. This "large and in charge" part of your brain will be part of the focus of our next chapter.

assemble different groups of neurons into teams, depending on your current goal or plan. We call these signals "top-down" processes, because their goal is to impose order on your thoughts, feelings, and actions.

You might think of the goals of these different kinds of signaling like two ways of putting together a puzzle. Your high-frequency, bottom-up signals try to figure out what the puzzle looks like by putting together the pieces that have similar colors and shapes and building it from the ground up. Your low-frequency, top-down signals look at the picture on the box and then decide where the pieces in front of them might belong based on this "idea" or "plan" for what the puzzle is supposed to look like.

Of course, this kicks the difficulty of your brain's version of Telephone up yet another notch. At any given time, the information from your inner and outer worlds is competing to drive your thoughts, feelings, and behaviors. And you've already learned what happens when low-frequency waves collide with higher-frequency ones. So how does this connect with what you've learned about your neural rhythms from the assessments?

The first thing to note is that your brain's preferred, low-frequency, wavelength for communication seems to correspond to the rate at which "packages" of information are carried to your inner world from your sensory neurons. As you've learned, most people's largest groups of rumbling, inner-world neurons cycle between thunderous sounds and silence at rates of about 7 to 14 cycles per second. And because their rumbles can easily drown out the smaller groups of chirping, outer-world neurons, they tend to function like gates. The chirps can best be "heard" in the moments of silence between the rumbles!

Perhaps the best evidence linking the speed of a person's alpha frequency to their outer-world sampling rate comes from a study by Roberto Cecere and colleagues. The researchers were interested in how different brains might understand continuously changing sensory information, based on the rate at which their alpha rhythms allow "packages" to come in from the outer world. To test this, the experimenters played around with a known visual illusion that occurs when a single flash of light is presented in sync with the first of two discrete auditory tones. On average, when the tones are separated from each other by about 100 milliseconds, many people think they see *two* flashes of light, even though only one is presented. Noting that this effect is strongest when the tones are presented every 100 milliseconds, which is the same rate as the most common alpha frequency (10 Hz), the researchers wondered whether people with different preferred alpha frequencies would experience the illusion with different degrees of spacing between the tones.

The reasoning behind this involves the collision of top-down and bottom-up information-processing streams. To be more specific, when two pieces of information are presented at roughly the same time—which we'll define as occurring within the same alpha window—your top-down, inner-world neurons are more likely to integrate them into a single mental object. You know the whole "If it looks like a duck and quacks like a duck" line of thinking? Because the first tone and the flash of light occur at the exact same time, your inner-world neurons interpret them as a single event happening in the world "out there."[15] The question then becomes: What does your brain do with the second tone? The idea here is that *if* the second sound is delivered in the same alpha package, your top-down neurons perceive them as part of the first event. They can accurately understand the stimulus as one flash of light that coincides with two noises. But when the second sound comes in a different alpha package all by itself, your inner-world neurons are more likely to "fill in the blank." Because they believe that the light and the sound are

15. Based on what you've learned so far, can you guess which *hemisphere* would be most likely to integrate information coming in from your eyes and your ears into a single event?

coming from a single source, they "assume" that there must have been two of these experiences and create the illusion of the second flash of light.

To test this hypothesis, Cecere and collaborators measured each participant's individual preferred alpha frequency from recordings of their brain at rest. Then they showed them a bunch of trials in which lights were flashed and tones were played and asked participants to report how many flashes of light they saw. This time, in the tricky "one light with two tones" conditions, they varied the amount of time between tones in precise 12-millisecond increments. As they predicted, their data showed a strong relation between each individual's alpha frequency and the amount of time between tones that would give rise to the illusion of a second flash. Specifically, because people with faster preferred alpha frequencies had faster sampling rates, they were more likely to see the illusion when the tones occurred closer together in time; whereas people with slower alpha frequencies were more likely to see the illusion when the tones occurred farther apart in time.

So why might this sampling rate relate to your working memory capacity? Though this is still a very active area of investigation, one possibility is that the rate at which your inner-world neurons can receive *new* information from the outside world is also related to the rate at which they can *refresh* the bits of information they are trying to hold on to. You might think of this like the mental juggling act that it is, in which the things you are trying to hold in your conscious awareness—like the numbers or letters—are the balls being juggled. The gravity that pulls the balls toward the ground is the process of forgetting, and as the juggling metaphor suggests—the contents of working memory are quickly "dropped" if they are not refreshed. The juggling metaphor also captures the fact that there are a finite number of "things" you can juggle, which is your capacity. And as you may have noticed during the test, sometimes, when you get one more number or letter than you can juggle, *all* of the previous ones seem to fall out of your memory, not just that extra one. Having a faster alpha frequency is like having faster hands—even if gravity stays constant, you can keep more balls in the air if you need less time to throw each one back up.

Though I admit that being able to juggle a lot of mental balls sounds appealing, one of the potential costs of having a faster sampling rate is that there would be less information traveling to the inner world in each package. Might people with slower alpha frequencies, with longer temporal windows of information bundled in each package, be able to make broader connections?

Research by Bazanova and Aftanas, who measured creativity using a task similar to the triangle test I gave you, suggests that this may be the case. In their study, preferred alpha frequencies were recorded in 98 individuals, who then completed a standardized creativity test. During the test, each person was given many different versions of the triangle task, in which they were asked to draw as many pictures as possible using a specific shape, in five-minute windows. Performance on the test was scored in terms of both *fluency*, or how many different pictures a person could come up with, on average, in the five-minute time period, and *originality*, or how rare any given response was. For instance, an image of a house with a triangle roof would be scored as less original than an image of a stegosaurus with triangle scales down its back. Their data showed that, in general, people with *faster* alpha frequencies produced *more* responses overall; but people with *slower* alpha frequencies produced *more original* responses, or were more creative.

In other words, speed isn't everything.

Once again, the moral of the story is that different doesn't necessarily equal better or worse. In the next section, we'll wrap up what you've learned about how your brain is orchestrated, and how this relates to your other engineering designs. With this information in mind, you'll be prepared to take your brain out on an adventure and explore the ways that *different* brains accomplish the fundamental jobs that they are tasked with.

Summary: Thinking at different wavelengths shapes the way we juggle information from the inner and outer worlds

The assessments in this chapter provided two complementary pieces of information about how your brain is orchestrated. The more items you were able to "juggle" in the working memory test, the faster your brain's preferred inner-world sampling rate is likely to be. But research on creativity suggests that people who can juggle lots of mental activities at once may produce many ideas, but fewer of them will be "outside of the box" when compared to those whose neural choirs have more "bass," so to speak. And those of you whose brains communicate more strongly in lower frequencies are also more likely to rely on insight when trying to solve problems than are those who rely less on these slower communication channels and are more likely to engage in deliberate problem-solving strategies.

Though we haven't talked much about how your brain becomes synchronized over a particular frequency, I'm going to sound like a broken record when I implicate both nature and nurture. And since the higher-frequency communication circuits primarily respond to the environment around you, you probably won't be surprised to learn that it's our low-frequency, inner-world communication systems that are most heritable. In fact, data collected on over 500 pairs of twins suggested that a whopping 81 percent of the variance in preferred alpha frequency is explained by genetic contributions! But as the large working-memory study by Clark and colleagues also showed, our preferred alpha frequencies change across our life spans.

From infancy, until it reaches its fastest rate around twenty years of age, the average alpha frequency will increase about 5.5 Hz, doubling in speed in the majority of us. Subsequent aging seems to decrease our tempos, though how much it slows is still a matter of debate, with estimates ranging from a slowing of .5 Hz–2.5 Hz by the age of seventy. I'm sure there are individual differences in this, and that they might relate to both genetics and the kind of lifestyle you live.

For example, mindfulness or meditation practices have been shown to influence neural synchronization in interesting ways. Of course, it's important to note that the nature of these practices varies quite a bit, but they tend to share a common focus on *internally* guiding your attention and awareness, as opposed to reflexively responding to thoughts or stimulation from the outer world. Neuroscientists have been recording the brain responses of trained and novice meditators for more than fifty years. This vast body of research overwhelmingly shows that when an individual is in a meditative state, their alpha frequencies become louder, or more *in sync*, and as a result, they are less distracted by intrusions from either the inner or outer world. Some data also suggest that during meditation, a person's sampling rate, or individual alpha frequency, also slows down.

But evidence about whether these practices produce *lasting* changes in your neural rhythms has been mixed. Though much research has shown that expert meditators have different patterns of brain activations from novices, the causal relation between meditation experience and brain function is difficult to determine. Are people with more in-sync inner-world neurons better able to learn how to meditate, or does meditation practice actually change a person's ability to control the flow of information into and out of the inner world? One of the few experiments that measured *changes* in brain activity within the same group of people as they learned to meditate showed that three months of intensive meditation training resulted in a significant slowing of their alpha frequency. Not all meditation studies report the same effects, however, which may suggest that different types of training lead to different neural consequences, or possibly that the type of meditation training one receives interacts with the way their individual brain prefers to process information. Nonetheless, the fact that neural synchronization changes *during* mindful exercises is much less controversial.

Externally driven experiences, like action video games, seem to shape neural communication at the other end of the frequency range. Some research suggests that gaming can increase people's peak alpha frequency ranges, at least temporarily. And if that isn't your cup of tea, might I suggest an *actual* cup of tea, or coffee? In fact, research has shown that consuming 250 milligrams of caffeine (about two cups of coffee) both increases the preferred frequency of your inner-world neurons and shifts the balance from inner-world toward outer-world neural synchronization.

To summarize, although a large percentage of your brain's preferred sampling rate is genetically determined, the kinds of things you ask your brain *to do* on a regular basis can also influence the speed with which your sensory neurons carry updated information to your inner world about what's going on around you. This has been a recurring theme in the first half of this book. You may have been born with a certain type of brain engineering, but the environment your brain exists in also has a significant effect on how it works. For example, people with the most lopsided brains seem to be able to develop specialized processing modules in their left hemispheres, while their right hemispheres try to take it all in and understand the "big picture." However, no matter *how* asymmetrical your neural computations might be, you need to have experience with a certain type of task or event before your brain can attempt the divide-and-conquer approach. Similarly, in "Mixology," we showed that genetic influences on dopamine or serotonin communication circuits became most apparent when people were placed in specific situations.

The differences in dopamine communication between extraverts and introverts, for instance, seem to be tied to how their brains respond to unexpected rewards, while the differences between people with higher or lower serotonin reuptake mechanisms seem most apparent when they are placed in stressful circumstances.

In the second half of this book, we're going to flip the nature/nurture problem on its head. Rather than focusing our attention on the biological designs of different brains, we're going to discuss how unique brains go about accomplishing the critical *jobs* that all humans need to accomplish to survive. In other words, if your brain were a car, we'd be moving from talking about whether you have four-wheel drive to discussing the different paths you might take to work. If your brain is a Honda Civic with excellent gas mileage and a good stereo system, you might be better off sitting in traffic for a bit and listening to your favorite podcast. On the other hand, if your brain is a Subaru Outback, and you can afford the gas and the new tires you might have to buy if you pop one, you might consider going "off-road." As soon as you're ready, let's take your brain out for a test drive in the real world.

PART 2

BRAIN FUNCTIONS

How Differently Engineered Brains Drive Us

My first memory is of the exact moment I realized that riding my tricycle down the stairs was a bad idea. Unfortunately for all involved, this didn't happen until *after* I had already begun said exercise. I do have a vague recollection of rolling my trike out of my bedroom and standing beside it atop the stairs. I must have paused there for a second to "think" about what I was doing, because I remember looking down at the carpeted stairway in front of me. In true first-person-toddler perspective, it seemed to stretch on forever.

Regrettably, at two and a half years old, I lacked both the life experience and the physics knowledge to realize that the ninety-degree turn at the bottom might create a problem. The next thing I remember—vividly—is that as I was approaching the wall at 100 toddler miles per hour, I had what was probably my first ever *Oh Shit* moment. Then, fade to black.[1]

From the vantage point that forty-six years of life experience provides, it seems entirely appropriate that the story of *me*, as recorded

---

1. Thankfully, I don't remember hitting the wall *or* the ground, though it must have hurt like hell because I broke my leg in the crash.

by my brain, begins with an epic fail. I certainly hope, for the sake of your caregivers at least, that the beginning of *your* story is a little less Evel Knievel. But what can I say? You live and you learn. And with a bit of luck, and a whole lot of brain processing, these things *typically* work out like that.

In fact, living and learning are at the heart of what the second half of this book is about. From the mundane activities, like estimating the angle your bike can turn at a given speed to the more profound decision points in life, like figuring out whether a particular choice will bring you more joy or more pain, your brain spends every waking moment engaged in elaborate problem-solving and decision-making algorithms. And of course, each brain goes about it a bit differently.

For example, one of your brain's most important jobs is to decide which of the thousands of pieces of information bombarding it at any given time are the most important. In the next chapter, "Focus," we'll discuss how the different types of brain designs we've discussed relate to the kinds of information that might capture their attention. As you'll learn, this has large implications for how your brain functions. It influences which part of an experience you're most likely to remember, what you'll learn from it, and whether you're likely to make a different choice in the future. Clearly, the fact that I still remember the stair-triking incident suggests my brain *hoped* this would be a "teachable moment" in my life.

Then, in "Adapt," we'll discuss what teachable moments really look like. Though many of you probably have a preconceived notion about what kind of learner you are, the vast majority of the things we learn in life are neither written in books nor taught in the classroom. In fact, *every* lived experience physically changes your brain, resulting in a brain that is fine-tuned for operating in the environments that shaped it. And I think you might be surprised to learn what *counts* as an experience from your brain's perspective. Ultimately, these experiences fundamentally shape the way we see the world we inhabit as well as how we understand people and situations that we don't have a lot of experience with.

At the end of the day, the reason your brain works so hard to adapt to its environment and learn what to focus on is so it can make better choices. As you'll learn in "Navigate," the dopamine circuits you were introduced to in "Mixology" play a particularly important role in helping people learn about the outcomes of their decisions. But parallel brain circuits drive decision-making in different ways, and certain outcomes have a bigger influence on some people than on others. For example, I can say with confidence after several, somewhat more successful, "stair adventures" that stairs *can* provide a variety of exciting opportunities. We'll also discuss how we use our memories to build maps of our knowledge of the world, and our place in it, both literally and figuratively. When we do so, we rely on our brain's powerful storytelling abilities to extract patterns and use those patterns to make inferences that link places and events in our experiences in meaningful ways.

But what happens when your navigation goes awry? What if the thing you thought would happen, based on your previous experiences, didn't happen? Or what if you find yourself in an environment you haven't adapted to, so you don't know what to expect? In "Explore," we'll discuss the brain processes that create curiosity and the behaviors that drive brains to try to update the gaps in their knowledge when they encounter them. But how does your brain decide whether there are dangers lurking in those unknown waters? In any truly unknown situation, there is always the potential to learn something new and useful, or to find something that could hurt you physically or psychologically. As you'll learn, there are both stable individual differences as well as situational influences that shape how much a brain is willing to take the risk and explore uncharted territory.

And last but definitely not least, we'll discuss one of the most important territories that we can *never* explore: the content of another person's mind. In "Connect," I'll describe the fundamentally different ways our brains try to understand others. After learning how strongly your brain design influences the way you understand the world, you might not be surprised to learn that social neuroscientists find *strong* evidence for *homophily* in the brain—the fact that we tend to hang out with other people whose brains work a lot like our own. As you'll learn, this is probably because our instinctual way of understanding one another is to see others as mirrors of ourselves. Of course, when you apply this tactic to a person whose brain *doesn't* work like yours, things can go wrong. For instance, because my mom is a practical person, there was *nothing* in her lived experience that prepared her for the fact that I would try to ride my tricycle down the stairs. Ironically, the reason she moved my tricycle upstairs into my bedroom with my other toys was because she was afraid someone would trip on it.

Oops.

But enough about me and my rather flat learning curve.

Now that *you* know something about *your* brain's design, it's time to take that baby out on the road and see if we can get some more insights about how it accomplishes these important daily functions. And whether we decide to take the stairs or not, it's bound to be a bumpy ride!

CHAPTER 4

FOCUS

How Signals Compete to Control Your Mind

If the subtitle of this chapter makes you feel a little nervous, trust me, you're not alone. When I tell people about my day job, questions about "mind control" and "reading people's minds" tend to follow. To be fair, I didn't make this any easier on myself when, in August of 2013, I hooked Andrea's brain up to the brain of our friend and colleague Rajesh Rao so they could play a video game together across campus. To be a bit more specific, with Andrea's consent (of course), I gave Rajesh's brain *control* of the part of Andrea's motor cortex that moves his hand. As Rajesh watched a video game from the computer science building on one side of campus, we recorded the electrical activity over the part of his brain that is responsible for controlling his right hand. Since *thinking* about moving your hand changes the balance of slow-wave communication to the faster frequencies used to perceive and interact with the world, our computer algorithm learned to detect when Rajesh *wanted* to move his hand.[1] And when that happened, the computer sent a signal across the Internet to our lab at the Institute for Learning and Brain Sciences on the other side of campus, which

---

1. OK, I admit it. That *is* a little bit of mind reading.

triggered our TMS machine. As you might remember from "Lop-sided," TMS is a machine that uses magnetic fields to induce little elec-trical currents in the brain. Because I had positioned the TMS coil over Andrea's left motor cortex, when Rajesh thought about moving his own right hand, Andrea's right hand moved. And because Andrea's hand was resting on the keyboard, Rajesh was essentially able to play the video game using Andrea as a highly overqualified joystick. Though the experience has been re-created in several fancy media ver-sions, if you're interested in watching it, I recommend finding the original footage on YouTube.[2] With this feat, we became the first team in the world to demonstrate that information could be transferred from one human brain to another directly. Unfortunately, we also scared the living shit out of a lot of people when we did.

But I am still reluctant to say that Rajesh was controlling Andrea's *mind*. My experience on the receiving end of a brain-to-brain inter-face is that the conscious experience feels much more like a reflex than like a thought inception. You're not even *aware* that your hand is moving until you feel it or hear the click of the keyboard it's push-ing. We are very far from being able to transfer the *desire* to push a button from one brain to another!

I realize that this may not do much to alleviate your concerns that I am an evil, mind-controlling scientist, and I don't blame you.[3] In fact, if you'd like to delve deeper into the ethical debates surrounding brain-interfacing technologies, I strongly recommend the documen-tary *I Am Human*, which discusses our work alongside other neural engineering technologies.[4] In the meantime, because I don't want

---

2. My main contribution to this video comes in the form of an off-camera laugh about 1 minute and 18 seconds in.

3. Those who know what a good partner Andrea is are probably even more con-vinced that I've bewitched him in some way!

4. The film also reflects the intellectual sophistication and heart of its brilliant directors, Taryn Southern and Elena Gaby.

the idea that I'm controlling my husband's mind to distract you from learning about how *you* control *your own*, please consider the following facts about human brain interfacing: First, *all* of the technology we have available for noninvasively putting information into a brain is very crude compared to the resolution of an actual thought. With a magnetic pulse, we can make your finger twitch, or even make you "see" a flash of light that isn't there. But we are much further away than most people imagine from being able to induce any sort of sophisticated perception, let alone being able to create *Inception*-like transmissions of thoughts. Second, it is *not possible* for any of these technologies to be used without a receiver's knowledge, or even without their consent. As you can see in the YouTube video, both Andrea and Rajesh are sitting very still, one under a cap that reads his brain waves and the other under a coil that has been carefully positioned over a particular spot on his head, with centimeter-level precision. Of course, I *could* use force to make someone participate in these experiments—but it would be a lot easier and more effective to just force them to do whatever nefarious things (rob a bank?) one might imagine I'm trying to get them to do in the first place. And this brings me to the third point, which is central to this chapter: How much do you understand about what *is* controlling your mind? Suffice to say that there are plenty of "mind control" signals in the environment all the time. And they have a stronger influence on you than I imagine more direct brain interfaces ever will. Whether you're watching a bikini-clad supermodel eat a messy hamburger during a Super Bowl commercial or reading about "Pizzagate" on the Internet, the words and images that bombard our brains the "old-fashioned" way have a massive influence on the way we think, both individually and collectively.

Though I completely understand why someone would feel uncomfortable about the idea that an external person, or message, might influence their thoughts (and by virtue of that, their behaviors), I am also at peace with the fact that most people have very little idea about what *is* controlling their minds. What does it even *mean* to be in control of your own mind? And what role does your conscious awareness play in the process? Questions like this have occupied philosophers' brains for millennia. And though neuroscientists have some catching up to do, we're learning more every day about the relation between different types of awareness and control. In the next section, we'll cover the basics about the different ways information can capture your focus and how this relates to mind control.

Understanding the relation between focus and mind control

The first thing I'd like to point out is that there are different *ways* that a piece of information might enter your conscious awareness and capture your focus of attention, and that they fall under a hierarchy of "mind control" situations. At the bottom of this hierarchy lie the processes involved in the reflexive, noticing type of focus. This is the space in which some piece of information *grabs* your attention, regardless of what you want to be doing at the time. Whether you're ruminating on some worry or turning your head to focus on a squirrel moving in your peripheral vision, this type of mind control occurs when your brain *automatically* weighs the signals being passed around in its game of Telephone and assigns them some sort of priority based on what it thinks is important.[5]

In the middle of the hierarchy is the place where more controlled, and more flexible, focusing happens. Here, the information you maintain in your working memory can be used to *guide* your noticing at the lower level. The paragraph-reading example in the "In Sync"

---

5. You'll learn more about how this works in the next chapter.

chapter shows how this type of focusing works. In one condition, your attention is captured by the meaning of the words. In another, it's focused on punctuation. And in the third, it's the sounds of words that capture your mind. This is the *costly* brain system where your conscious thoughts can be used to override your more automatic ways of prioritizing information. And if someone is asking you to *pay attention*, this is what they are asking your brain to do.

Finally, at the top of the hierarchy of focusing are the processes involved in self-awareness. This is the place where the focus of our "mind's eye" points inward and tries to assess whether or not the *way* we're doing the things we're doing is moving us closer to our intended goals. Here, your brain tries to answer questions like "Did I study long enough to get an A on that test?" or "Why do I keep losing my temper in this circumstance?" by taking a read on the parts of its own processing that you have conscious awareness of.

Though there are different computations that underpin each of these types of focusing, they all share a common constraint: We can only be truly *aware* of very few things at one time.[6] Regardless of how the information gets "in there," that is, into that place in your conscious mind that allows you to say, "I'm thinking about X," once it's there, something *else* gets pushed out. So the things we notice reflexively, our attempts to control our noticing, and our mental navel-gazing processes all compete for access to that limited-capacity conscious workspace. And of course, the extent to which *your* conscious thoughts are captivated by these different types of focusing will vary depending on your brain design and the way your life experiences have shaped your brain. In the next section, we'll talk a bit

---

6. The precise number of things you can think about at once really depends on what you count as a "thing," what you count as "thinking," and how you define "at once." If you define a "thing" as something you can pull out of a context and manipulate in some way, a "thought" as something that requires conscious awareness and can execute control over other processes, and "at once" as occurring in the exact same instance, the number is somewhere between one and four things.

about what we have learned about the relation between brain design and the extent to which these different types of "mind control" can influence an individual's thoughts.

Lopsided focusing

It seems appropriate to start our discussion of the way different brains function the same way we started our discussion of how different brains are engineered—with a tale of the two hemispheres. As it turns out, the different computations that characterize the left and right hemispheres of typically lateralized brains also give rise to very different ways of focusing. This difference is strongly apparent in the family of brain conditions characterized by "neglect," or the inability to *notice* something despite having completely intact sensory systems for detecting that thing. The most common type of neglect observed in patients with brain injury is "hemispatial," which typically happens following injury to the *right* hemisphere. At risk of oversimplifying this condition, the competition for focus in people with hemispatial neglect has been dramatically reduced, simply because their brains don't *consider* information from half of the outer world. For example, if you ask a person with perfect visual acuity who has damage in their right parietal lobe—the part of the brain that sits on top and in front of the visual cortex and connects it to the frontal lobes—to describe what's happening in the world around them, they will almost exclusively describe things to the right of their noses. If you set a plate of food in front of them, they might only *eat* what is on the right half of the plate. If you ask them to copy a picture, they will draw the right half of it. And in one of the most interesting "made for TV" demonstrations of this condition, if you ask them to draw a clock from memory, they will crowd the numbers around the right side of the clock's face, leaving the left half blank.

But these failures to notice aren't only limited to the things a person with neglect *sees*. If you ask them to show you how they would execute an everyday task like shaving, or combing their hair, they will often mimic the action over only the right side of their body. Sometimes they even *forget to dress* the left side of their bodies. And one of the most remarkable things about this condition is that the patients don't notice what they don't notice! Unlike those with comparable deficits following left-hemisphere damage, patients with injury to their right hemisphere are much more likely to exhibit anosognosia, or the "lack of ability to perceive the realities of one's own condition." Their challenges with focusing ascend all the way from reflexive up to self-awareness. And though sometimes ignorance really is bliss, this kind of failure to notice can have devastating implications on how likely someone is to seek out, and respond to, treatment.

By comparison, damage to the left hemisphere rarely results in any type of neglect syndrome. And though this hasn't been studied systematically, to the best of my knowledge, it's entirely plausible that those who do experience deficits in attention following left-hemisphere damage have less typical, or more balanced, brain lateralization. But even in these cases, research has shown that patients with left-hemisphere damage are *aware* of their challenges, which makes it easier for them to learn strategies that can compensate for their limitations in focusing.[7]

Based on the dramatic differences in attentional deficits that occur with damage to the left versus right hemisphere, many researchers have proposed that in healthy brains, one hemisphere (typically the left) drives the more controlled, goal-directed type of focusing, while the other (typically the right) is governed by the more

---

7. One example of this is that patients are trained to rotate their plates 180 degrees when they think they've finished their food. Then—like magic—a new plate of food seems to appear!

automatic ways of figuring out what to focus on. And this asymmetry makes sense, in light of what we've discussed about the different computational specialties of the left and right hemispheres. At least in traditionally lopsided brains, the left hemisphere's tendency to execute many fast, specialized processes in parallel makes it well suited for selecting a particular stream of information to amplify. Meanwhile, the right hemisphere's ability to integrate lots of different streams of information into a connected pattern makes it particularly good at *noticing* when something seems unusual or out of place.

This division of types of focus in the two hemispheres is also consistent with the functional goals of the two hemispheres proposed by Dien in the Janus model. As a reminder, the model proposes that the left hemisphere's goal is to anticipate the future, while the right hemisphere is focused on the here and now. It makes sense that a future-focused left hemisphere would be using goals and plans to direct its spotlight of attention on the pieces of information it deemed most *relevant* for predicting outcomes, while a right hemisphere, with the goal of understanding what's happening right now, would try to notice the world as it is.

But here's a question we only touched on briefly in "Lopsided": What happens when the pieces of information capturing the attention in your left and right hemispheres are different? Remember the story of Vicki, the callosotomy patient who had to wrestle with her left hand to get the item she wanted out of the closet? In the extreme case of two hemispheres with severed communication, Vicki seemed to only be *aware of* the motivations of her right hand (driven by the left hemisphere). I originally related this to what Gazzaniga called the "interpreter" function, by which the left hemisphere (in most people) constructs the story of *why* things are happening. But now we can add another piece to the puzzle—the idea that the left hemisphere's monologue about what's going on in the world *drives* the types of information it's paying attention to, while the right hemisphere (in most people) is more likely to respond automatically to what's happening in the environment around it.

Suppose Vicki had a plan to wear a pantsuit and practical shoes to work on Monday, because she knew she had a lot of walking to do that day. Her right hand would be responding to the things her left hemisphere found that fit that category. But what if, along the way, her right hemisphere saw a cute purple dress that grabbed its attention? In connected brains, that information would have a chance to compete for her attention and subsequent decision-making. Critically for understanding how *your* brain functions, an increasing body of research suggests that differences in how easily your mind can be captivated, or distracted, by your automatic noticing processes versus driven by your goal-directed thoughts, relates to how the signals from your two hemispheres compete for attention.[8] In the next section, we'll do a quick assessment to figure out where you lie in this focusing space.

Assessing your focus

You've already got a pretty good idea of how lopsided your brain is, but there's an easy test you can do at home to measure how your two hemispheres contribute to focus more specifically. You'll need a pencil, a piece of paper, and a ruler or measuring tape. In the book, I've given you an example of a series of horizontal lines, drawn in

---

8. As I typed the word "distracted" in this sentence, I remembered that I needed to charge my headphones so I can listen to a book (*Driven to Distraction* by Edward Hallowell and John Ratey) on my dog walk this afternoon. My headphones were in my bedroom, and when I went in there to get them, I noticed my pajamas were still on the floor, so I picked them up and put them in with the dirty clothes. My dog Coccolina followed me into the bedroom and shook her head, which reminded me that I needed to clean her ears—so I went into the bathroom to get some Q-tips and decided that my teeth felt hairy, so I brushed my teeth. Fortunately, after this ten-minute detour, I *did* remember to charge my headphones, and also that I am in the middle of writing a book. This is what it looks like in my brain when the two hemispheres wrestle for control.

different positions on the page. If you use these, you won't need a piece of paper. But if you don't want to draw on your book, you can re-create the experiment at home by drawing ten or so lines of different lengths that are not perfectly centered on a piece of paper.

Your goal is deceptively simple. Without using anything to measure them, take a pen or pencil, and mark a vertical line to indicate your best estimate of where the *center* of each line is. Ready to get started?

LINE BISECTION TASK

There are different levels of precision you can use to assess yourself—depending on how much time you want to spend and how accurate you want to be. The quick-and-dirty way to do this is just to count the number of times you marked your line to the left versus the right of true center and see whether you systematically erred in one direction or the other. To do this, you don't really even need a ruler. Just get another piece of paper and make a mark the same length as the first half of your line, then drag it to the second half. If they're the same size—you got the dead center. If the second half is shorter, you marked your center to the right of true center, and if it's longer, you marked your center to the left of true center. Using this method, you can get an idea of how lopsided your focus is by figuring out the proportion of times (out of ten) that you erred in the most frequent direction. For example, if you marked equally frequently to the left or right of true center, you'd get a 5/10 and have a very balanced way of focusing. But if you marked to the right of true center on all but one line, you'd score 9/10 and have a pretty lopsided pattern of focusing.

If you prefer, you can get a ruler and be even more precise by measuring the distance you are off from true center on each line. If you mark left of true center, assign that distance a negative value, and if you mark right of true center, give it a positive value. Then add all ten distances together and divide them by ten. This will tell you how lopsided your focus is, on average. For instance, you might find that across ten trials, you were within one-eighth of an inch away from true center—this would reflect a pretty balanced type of focusing. The greater the distance you were away from true center on average in either direction, the larger your asymmetry in focusing is.

So which of your hemispheres is driving your noticing?

Most *typical* left-hemisphere-dominant people will mark the line to the *left* of its true center more frequently than they will to the right. The more consistently you marked the lines to the left of their true centers, the more likely you are to be driven by left-dominant, goal-directed attention processes. By contrast, people with more balanced or right-dominant brains may have marked the lines to the right of their true center more frequently. This pattern of responding has been linked to more distractable, or "organically driven" focusing.

In fact, people with attention deficit/hyperactivity disorder (ADHD) regularly bisect lines to the right of their true center.[9] This is one piece of a growing body of evidence suggesting that the symptoms of ADHD are at least partially related to the competition for focus between left and right hemispheres. Consistent with this view, ADHD is diagnosed more frequently in non-right-handed individuals! It's important to keep in mind, however, that ADHD is not *actually* an impairment in attention as a whole. Instead, it is better characterized as a pattern of focusing that is more influenced, on average, by automatic noticing processes than it is by the more controlled "pay attention" type of focusing mechanisms. This is not to say that people with ADHD can't pay attention—it's just that for them, the *cost* of doing so is much higher. To better understand how this works, we need to dig a bit deeper into the idea of mind control. What does this process of grappling for control of your conscious awareness look like in the brain?

The rhythms of mind control

The funny thing about *mind control* is that, at the end of the day, most scientists, educators, and parents assume that more is better—at least as long as the mind in question is controlling itself. Because of this, the majority of the research on differences in how automatic and controlled processes interact is focused on the controlled side of the equation. Viewed under such a lens, it is easy to overlook the fact that some brains have to work *harder* to achieve control, or to understand what the benefits of more spontaneous focusing may be. And yet I

---

9. This certainly does not mean that bisecting the lines to the right of their centers is *diagnostic* of ADHD. But as I mentioned in "Introductions," the symptoms of ADHD fall on a continuum, and you may be more likely to experience some of the symptoms.

think everyone has had a moment in which they were *trying* to do some mental activity—whether it was as simple as remembering a name or something more complicated, like trying to solve one of the problems in the last chapter—and you know, instinctively, that the only way to find the answer is to stop *trying* to. *It'll come to me,* you might've thought, having learned the hard way that shutting the controlled part of your brain the hell up is sometimes the best thing for you. This is because we *need* both types of attention, and there are times when too much of one or the other of them can get in the way.

I like to think of the relation between automatic and controlled attention mechanisms like the partnership between a horse and a rider.[10] The horse is the automatic noticing part of your brain. It has learned, through its experiences and instincts, what it needs to focus on. Without any guidance from the rider, the horse will choose the best place to put its feet. It has a strong survival instinct, and if left alone, the horse will move toward good things, and avoid bad things. And if it comes across something new, or even an old thing in a new place, the horse will stop and take a good look before deciding what to do next. This last bit may annoy the rider, but eventually, they're still going to get where they want to go faster, on average, if they're riding a horse than if they're walking. And the horse's way of noticing is precisely the way behaving species have survived on Earth for *hundreds of millions* of years. It is a type of focus grounded in the present,

---

10. I had the idea for this metaphor deep in the middle of the me-search involved in writing this book. One morning I said to Andrea—"I feel like my brain is a horse and I can't figure out whether I need to take a stick to it or pat it on the neck for reassurance!" As a girl who spent the first thirty years of her life wanting a horse of her own, and the next fifteen trying to figure out how to control the one I got, I *really* became invested in this metaphor. But alas, I wasn't its inventor. The metaphor of the horse and rider (or elephant and rider) has been used by many—from Sigmund Freud to Tim Shallice—to explain some aspect of the human mind.

used to respond quickly and efficiently to one's environmental surroundings.

The rider, on the other hand, is the more controlled, "pay attention" part of your focus. They can be motivated by abstract goals that have nothing to do with their current surroundings and can use their riding aids to steer the horse in a way that will help them accomplish those goals. Hell, they can even use Google Maps to get themselves there. In fact, a skilled rider can get a horse to do things that the horse would have never *dreamed* of doing on its own, from herding a cow to riding into battle.

To extend the metaphor to our brain's mechanisms, anyone lucky enough to have spent time on the back of a horse knows that good riding requires a sense of timing. Because horses move in naturally rhythmic ways, there are periodic opportunities to either maximally or minimally influence what they will do next. For instance, if you try to turn a horse at the moment of a gallop when they have no feet on the ground, you're going to have a hell of a time. As you probably remember from the "In Sync" chapter, our brains are also sensitive to timing like this. In fact, the "aids" that the control center in your brain uses to shape your more automatic noticing processes are the neural oscillations that are generated by lower frequencies, like alpha. So when the "rider" part of your brain wants to use its goals to turn *down* the volume on the automatic noticing mechanisms they think are irrelevant, they turn *up* the volume of alpha in those regions. And as you'll remember, when lower-frequency rumbles from the inner world collide with the higher-frequency chirps from the outer world, the chirps can get drowned out. Of course, the inverse is also true. If the rider of your brain wants to focus on a particular piece of information that might not be intrinsically interesting to the horse (or the automatic processes), it can turn *down* the alpha power in a certain brain region and *increase* the volume of signals coming from the outer world. In essence, this is like giving that particular stream of information a head start in the competition for noticing. When this works perfectly well, the inner-world rider is "in control" of what types of information enter your conscious awareness.

One study by Saskia Haegens and colleagues demonstrated how this works using a tactile discrimination task. During the task, participants received bursts of electrical stimulation on either their left or right thumb. Their job was to decide whether they had received a fast burst (between 41 and 66 Hz) or a slower burst (between 25 and 33 Hz). The entire sensation lasted only a quarter of a second, and the strength of the stimulation was just above their established threshold for detection. To make the task even harder, at the beginning of each trial, both thumbs received some sort of stimulation, but the participants were instructed to ignore one hand and report *only* what kind of stimulation they received on the other hand. Before each trial, participants were given a cue that instructed them which hand to "pay attention to" next. Following the cue, the researchers measured alpha power in the areas of the motor cortex corresponding to each hand. They showed that, on average, alpha increased over the area corresponding to the to-be-ignored hand and decreased over the area corresponding to the hand of interest. Critically, the degree to which this happened on any individual trial predicted whether the participant would get the answer to the question correct. In other words, people were better able to *feel* the difference between the two sensations when the noticing was "turned up" on the intended hand by the decrease in alpha *and* "turned down" on the irrelevant hand by the increase in alpha.

Another study, by Rebecca Compton and colleagues, used a similar measure of the volume of communication in alpha frequencies to investigate how people perform on the Stroop Task—the color-naming test that I described in "In Sync" as a way of illustrating how automatic reading becomes. As you probably remember, because reading words is highly automatic for most of us, answering "BLACK" to a trial in which the word "RED" is printed in black ink is like trying to turn the horse part of your brain to the left while someone is standing with a bucket of sugar cubes to your right. Not surprisingly, the researchers found that as participants answered these conflict trials, alpha power increased, particularly over the right, noticing hemisphere. The people who found the task the *easiest*, as shown by the fastest response times, also had the biggest difference between alpha in the left and right hemispheres. More "silencing" alpha in the right hemisphere suggests that their "goal-oriented" left hemispheres were taking in information, while their right hemispheres were inhibited! And *this* is how mind control was achieved.

These findings, which link laterality and neural synchronization to differences in focusing, are also consistent with the study we discussed in "In Sync" that was conducted by Brian Erickson and colleagues. As you may recall, they found that greater alpha in the left hemisphere was associated with a higher probability of solving word puzzles through insight experiences, while lower alpha power in the left hemisphere was associated with more controlled, or systematic, searches for solutions.

Taken together, these studies show how differences in the computations of your left and right hemisphere interact with your brain's neural orchestration to influence what will capture the focus of your attention. You might think of this combination as an explanation for how *frequently* the rider of your brain guides the horse, or how likely they are to relinquish control and let their intuitive horses take them where they need to go. But one major problem with this metaphor is that it doesn't explain how the rider decides where to take your horse. If you assume that the rider has its own *brain*, you have to wonder whether that brain also has a horse and rider? If not, what's in control of *it*? This kind of "turtles all the way down" recursive thinking leads you on a path to nowhere fast. In the next section, we'll get into the nitty-gritty details of mind control, which includes a discussion of where it all starts and stops when the same biological system that gives you a *sense* of being in control is also the one calling the shots.

What's really controlling your brain?

Let me start with a bit of a trigger warning. "The Question" that functions as the title of this section is arguably one of the most important questions people ask about their brains—and they do, in contexts that range from spirituality through consciousness to volition. The answer, however, or at least our current *scientific* understanding of it, can make most people feel pretty uncomfortable.[11]

But we've done a fair amount of setting the stage for this in the earlier sections of the book. For example, in "Lopsided," we talked about the disturbing things that can happen when a divided brain gets the "feeling" that it is not in control of what one half of its body is doing, or of the way information coming in from one half of the world is influencing it. We also talked about how the left hemisphere will automatically make up stories that incorporate the things it doesn't understand and those it does into a connected explanation of *why* the whole human it's driving around is behaving the way it behaves.

Then, in "In Sync," and again in this chapter, we discussed how neurons that communicate in lower frequencies can be used to coordinate your mental activities flexibly, based on your current goals or intentions. But words like "coordinate," "guide," and "influence" make it sound a bit like some part of your brain has independent ideas that it uses to control the other parts . . .

---

11. If you are interested, I highly recommend reading Robert Sapolsky's discussion of this problem space in *Behave*.

Which brings us back to the whole brain-within-a-brain challenge. And though I've tried to be very honest with you about how much we don't know about how brains work, if I leave you with a little black box in the middle of your brain with a question mark that says "and then your brain decides what to do," I haven't really explained anything![12]

Fortunately (or unfortunately) for you,[13] though I've been a bit cagey about *who* is deciding *what* in your brain up until this point, it's not because I don't care about the details. On the contrary, I probably care *way too much* about them. Not only are they central to all of my biggest contributions to science, they play a critical role in the love story that united Andrea and me personally and professionally. This area of neuroscience is *so important*, and yet *so complicated*, it took me about three years to feel like I understood it well enough to write anything about it. And I really want to make sure to get it right in this book, because, as you've undoubtedly noticed by now, there are a lot of moving parts (86 billion, give or take). So please bear with me as I introduce you to the *basal ganglia nuclei*, the puppet masters of your brain and my heart, through a love story in three acts.

Act 1: I've been invited to coffee, ostensibly to discuss research, by a tall, raven-haired, Italian man. We'd both completed our doctoral degrees a couple of years previously, and we were working in different labs at Carnegie Mellon that use computational models to understand the mind and brain. During our first "date," I quickly swung the conversation into hard-core science.

---

12. This problem is called "the homunculus argument," and cognitive neuroscientists (including me) can fall into these kinds of explanations pretty easily! Once, over too many beers, I frustrated the crap out of my friend Rob (another neuroscientist) by refusing to accept the idea that "my brain decides" was a satisfactory explanation of how things work. This probably explains why I don't have very many friends...

13. I'll let you decide after sorting through the extensive details that follow.

ME [*because I'm not very good at casual conversations*]: So ACT-R[14] works by using a bunch of "If X, then Y" statements, right?

ANDREA [*excited because I know something about models*]: Yes!

ME [*because I'm even worse at flirting*]: But brains *don't* work like that.

ANDREA [*even more excited to argue with me about science*]: Well, actually,[15] the model I'm working on right now shows how I think one part of the brain does exactly that!

ME [*swooning*]: Tell me more.

Take-home message: Based on Andrea's model (which is also based on a lot of empirical data), the basal ganglia are a set of nuclei that can use information about *context* (for example, "If I'm in my own home as opposed to some strange place") to decide[16] which type of signals are relevant for a particular task (for example, "Then I don't have to pay attention to where I am in physical space" because I know my way around my own house very well). This is also really important for the *flexibility* we talked about in "In Sync," because the precise signal that's important in one instance might be the exact thing that you need to ignore in another. In essence, this process enables the "programmability" that comes with instructed, or goal-directed, behavior.

Act 2: Andrea and I have been dating for several months now. He is sitting on one side of my kitchen table, working on his model of the basal ganglia. During our time together, we have gotten into the habit

---

14. ACT-R is one of the widely used, if not the most widely used, computational architectures for modeling the mind, created by Andrea's postdoctoral mentor and friend, the brilliant John Anderson.

15. This word has an extra syllable when Andrea says it, which is wonderful.

16. I promise I will talk about *how* the basal ganglia decide at the end of this section!

of calling his model "The Baby," although its legal name is "Conditional Routing Model." I am sitting on the other side of the table, trying to understand the results of a study I've just run comparing the brain responses of people with different working memory capacities as they read sentences in the MRI scanner under a variety of conditions. I'm flummoxed, because while some of the results lie in cortical processing centers that make sense to me, this little area *in the middle* of the brain, the *caudate nucleus*, is also showing up as one of the things that people with high working memory use differently when they *read*. To be perfectly honest, like many people who study the complexities of human cognition, I'd learned that all of the important things happen in the cortex, on the outside of the brain, and not in those medial, "lizard brain" regions, for Pete's sake! So when I look up research on the caudate nucleus to try to figure out what it does, and I learn that it's part of the basal ganglia,[17] I get really excited! I'm hoping Andrea might help me understand what this thing is doing in my high-capacity readers. And it turns out that not only does his model give me some new ideas about why people with bigger working memory capacities read differently from people with smaller working memory capacities—one that is more about controlling attention than it is about how much stuff someone can hold in mind—it also explains the relation between the findings in the caudate nucleus and the cortical regions in the frontal lobe I was focused on.

Take-home message: Almost everything you read about brains will put the "rider" of your brain in the prefrontal cortex. And that

---

17. The basal ganglia is one (plural) name to describe eight anatomically separate brain regions. I *know*, it confused me at first too! But they function as a whole to perform the critical, signal-routing mechanisms we're going to talk about in a second. It gets worse, because smaller groups of these regions are collectively called by other names, like the *dorsal* and *ventral striatum*. The details of this are not really important, unless you're going to go read other, more technical, brain stuff, in which case you should know that the brain regions that make up the basal ganglia go by *a lot* of different names!

idea isn't totally wrong—but it's also not entirely correct. I'll admit, it's very tempting to give the prefrontal cortex all of the credit for our most sophisticated behaviors. It's the big, flashy, photogenic brain region that most strikingly separates us from chimps, after all. And it seems unlikely that the massive number of evolutionarily new gray-matter neurons that can be found there aren't doing something impressive. As someone who studies language, I have no interest in disputing this fact—I only want to add that the prefrontal cortex has an older and more experienced assistant that is critical to its job: the basal ganglia nuclei, which reside in the center of the brain.

The simplest way to explain this collaboration is that the prefrontal cortex holds the goal for behaving in mind, the "If" part of the equation, while the basal ganglia help with the execution of the "then" part. They provide the mechanism for turning up and down the relevant signals given the goal at hand. In short, the basal ganglia work behind the scenes to influence the information that *arrives* in the prefrontal cortex. In doing so, they function a lot like the algorithms the social media companies use to decide not only which of your friends' posts you want to see but also which advertisements and news stories to put in front of you.[18]

Act 3: Andrea and I have been married for over a year and are now doing some collaborative research at the University of Washington, investigating whether bilingualism has any effect on the basal ganglia's signal-routing mechanisms.[19] I've been asked to review a paper on the neural basis of autism spectrum disorder (ASD), not because I have any specific expertise with ASD but because the paper is

---

18. I'm not a fan of this, by the way, for reasons that will hopefully become clearer in subsequent chapters.

19. Incidentally, Andrea speaks three languages fluently. English is his third language, and he speaks it better than I do, so this is a vulnerable space for me. Apparently, the saying "what you can't do, you teach" is also true for research!

looking at neural synchronization, which I *do* have considerable experience with. But as I'm reading the paper, I start to realize how many of the characteristic behavioral patterns of ASD appear to be the mirror opposites of the behaviors we're finding in bilinguals. It's *really early*, and it's a Sunday morning, so I do a bit of digging before I climb back into bed and whisper, "Andrea . . . I think the basal ganglia is doing something *different* in autism." Though he is not an early riser, nor someone who appreciates being woken up early, he opens one eye and says, "Tell me more."

To better appreciate the "insight" moment I had that morning, it's important to know that many people who study the basal ganglia are focused specifically on motor control. This is undoubtedly one of the basal ganglia's evolutionarily oldest jobs, because motor control—like so many other controlled things—is driven by frontal cortex computations. And that morning I found out that abnormalities in basal ganglia size and function had *already* been documented in ASD, but they had primarily been linked to one of the symptoms—repetitive, or stereotyped, motor behaviors. What I realized the researchers in the field were missing was an understanding of how the "if-then" flexible computations described in The Baby were also relevant for language and social functioning, the other two classes of symptoms that are typically impaired in ASD.[20] Also, I thought that the irregular patterns of neural synchronization reported in the paper I was reviewing might be explained by irregularities in the basal ganglia's signal-routing mechanisms. So I wondered aloud, to Andrea's single opened eye, whether the basal ganglia in people with

---

20. To the best of my knowledge, as of 2020, language impairments are no longer a required part of the ASD spectrum; however, many individuals at the more severe end of the spectrum have language deficits, and all of them have some kind of social impairment.

ASD might *not* be functioning to flexibly turn the signals up or down based on shifts in goals.

Fortunately, through the network of researchers at the university, Andrea and I were able to find Natalia Kleinhans, an amazing clinical psychologist and ASD researcher, to help us explore these ideas. To do so, we analyzed functional MRI data from 16 adults with ASD and 17 age-and-IQ-matched adults without ASD diagnoses as they completed a task designed to measure attentional control called a "Go/No-Go" task. The Go/No-Go task is pretty boring, but it's a good way to look at the intersection between thought control and motor control. In Go blocks, participants press a button every time they see anything presented on the screen (in this case it was either a face or a letter, depending on the block). In more difficult Go/No-Go blocks, participants are told *not* to press the button if they see a certain kind of stimulus (for example, an *X* in the letter blocks or a sad face in the face blocks). In our experiment, half of the stimuli were No-Go trials. Based on what Andrea and I (as formally proposed in The Baby) believed the basal ganglia were doing in this task, we proposed that during Go/No-Go trials, basal ganglia activation should *decrease* the flow of signals between the occipital lobe, which is processing information about the stimulus, and the frontal lobes, which are both holding the information about the goal of the task and pushing the button—as evidence of attentional filtering. What we found was that in controls without an ASD diagnosis, this was indeed the case. However, in individuals with ASD, basal ganglia activation caused connectivity between the occipital lobe and the frontal lobe to *increase*. It was as if the basal ganglia were turning up the volume on *everything* in individuals with ASD.

Take-home message: The basal ganglia's job of turning less-important (or distracting) signals down is *at least* as important as the process of turning the relevant signals up. This is consistent with the sensory discrimination results of Haegens and colleagues that we discussed earlier in this chapter. People were better able to detect the rate of stimulation on one of their thumbs when their alpha rhythms turned *down* the signals coming from the other, simultaneously stimulated thumb. You can probably intuit why this is true by imagining that, while trying to read this book, you were unable to prevent yourself from noticing everything else that is going on around you—from the way your neck muscles are straining to keep that ten-pound head of yours upright to the pace of your breath to the color of the lighting in the room around you to the smells in the air to whatever cute, distracting, or outright annoying thing your roommate, pets, children, or plants might be doing right now—forget it. One of the important *points* of this chapter is that, in any given moment, the number of things that are happening around you that are unimportant for your decision space far exceeds the number of things that are important. And here's where our basal ganglia love story *finally* comes full circle and connects back to what you already know about how you work—each time the basal ganglia send a packet of modified signals to the prefrontal cortex, they get dopamine *feedback* about the outcomes of the decision that was made based on this type of signal routing. In this way, the basal ganglia can use dopamine to learn over time which types of signals to turn up and which to turn down!

Summary: The intuitive "horse" and the controlling "rider" compete to capture your conscious awareness with different types of information

To rehash this chapter's main topics, the basal ganglia nuclei lie not only at the center of my love story but also at the center of your brain, physically and metaphorically. Your basal ganglia nuclei are, for all practical purposes, the conductor of signals in your brain. Being in the center of everything perfectly positions the basal ganglia to have a great sense of the "state of the world" according to the gossip happening all around them. And they do. In fact, the basal ganglia are surrounded by white-matter fibers that carry high-speed signals in from practically *every* other region of the brain. These signals include sensory information about the world "out there" as well as information held in working memory about your current goals—the "if" that tells them what to do next. Laterality research also suggests that the goals may be primarily used to drive left-hemisphere focusing, while the sensory information from the world "out there" may be the center of an older, more intuitive, right-hemisphere way of noticing. As these signals converge on the basal ganglia nuclei, they use the previous dopamine feedback signals to decide which pieces of information are most important in the current context—the "then" part of this equation.[21]

This function is important because it allows the basal ganglia to "weigh in" on the massively overlapping signals arriving in the prefrontal cortex, and *bias* their processes based on what they believe your goal is. From there, the frontal lobes gain control—using lower-frequency brain waves to generate patterns of activation that will create a thought, a behavior, or some combination of the two. Then, as we touched on in "Mixology," the basal ganglia will use dopamine reward signals to determine if the outcome of whatever the frontal lobe's choice was, was better than expected, worse than expected, or exactly as expected. And this will shape the type of information the basal ganglia will route in the future, given the same goal. And so, at the end of the day, it's a very inanimate series of computations that controls you—a representation of your context or goals, a system for

---

21. You'll learn more about the role of dopamine and this feedback loop in "Navigate."

prioritizing or weighing signals, and another one for reprioritizing signals based on the outcome of what happened.

Isn't it romantic?

In the next two chapters, you'll learn more about the basal ganglia's learning process. How do your experiences shape the way your basal ganglia and cortical computation centers collaborate to figure out what's happening in your environment, and what to do about it?

CHAPTER 5

ADAPT

How Your Brain Learns to Understand
the Environment You Inhabit

As I put the finishing touches on this chapter, I am gearing up for the return to in-person teaching and research at my university after eighteen months of pandemic restrictions. I remember all too well that when we first received the "stay at home" order, my attitude quickly shifted from feeling like I got a "snow day" to feeling like a bird in a cage. Being a person who is both extraverted by nature and *not* one who enjoys being told what to do,[1] the transition was pretty rough.

But you know what?

I got used to it. And there were even parts of it I learned to *like*—like the convenience of being able to wear comfortable pants every day and having my dogs lying on my feet during meetings. And now that things have been deemed "safe enough" to go *back* into the office a few days a week, I find it *exhausting* to be around people in real life.[2]

Though the pandemic undoubtedly affected some of your lives

---

1. Does anyone? Of course, I typically *follow* the rules and restrictions when I know they have important health and safety consequences, and because functioning in a society depends on such things . . .

2. Plus, my dogs are seriously neurotic when I leave them now.

more than others, I know it touched every one of us in ways we will never forget.[3] And now, after a year and a half and counting, I am *positive* that every person reading this book is fundamentally different from when the pandemic started.

Because that's how our brains work. They are molded by our experiences so that we can *fit* into all kinds of different situations— even the decidedly suboptimal ones.

This is actually one of the most *human* things about all of our brains. In fact, according to some contemporary views of human evolution, our ancestors underwent a "cognitive revolution" precisely because they were forced to adapt. Based on evidence suggesting that the size of our ancestors' brains increased following periods of extreme weather instability, one popular explanation for our remarkable flexibility is that the hominids who were *not* able to adapt to environmental changes didn't survive. But those who could think and respond flexibly were better able to adopt new behaviors when the environment shifted from being stable and hospitable to becoming cooler, less predictable, and less forgiving. In other words, the brains of modern humans were *selected* for their ability to learn and adapt to changing environments.

Then, as thousands of generations of fast learners and flexible thinkers reproduced with one another, the brain and skull sizes of their offspring also increased. One major cost of this was that childbirth became more dangerous. And as a result, moms began giving birth to babies at developmentally earlier stages.[4] In fact, modern human infants, with brains that are only 27 percent of their adult

---

3. Sending my sincerest appreciation to those of you who had to risk your lives and keep showing up to work for the rest of us. And for those who lost someone close—my most heartfelt condolences.

4. I should point out that there are some pretty convincing arguments that human babies are born less developed because of the advantages this provides in adaptability—not because of their increasing head sizes—but to the best of my knowledge this is still debated.

size, are even *less* ready to hit the ground running than our prehistoric ancestors were. They don't even develop enough strength and coordination to support their own heads until sometime between the third and sixth months of life![5]

Although human infants are born weaker and more vulnerable than many other animals, their shortcomings are partially compensated for by their incredible ability to learn. In the place of hardwired instincts, we come pre-programmed without much more than a set of powerful learning mechanisms that allow us to adapt to dramatically different environments. As a result, humans now cover the planet, thriving in all habitats. And though I don't *know* if it will happen in my lifetime, I can readily imagine a future in which a human baby, raised on Mars with 38 percent of Earth's gravitational pull, could have a much more successful stair-triking experience.

But one of the major *costs* of this remarkable flexibility is that humans are born without any significant preconceived notions about how things work. William James, an American philosopher who helped to define the field of psychology, poetically described what it must be like to be born in such a state, in *The Principles of Psychology*.

"The baby, assailed by eyes, ears, nose, skin, and entrails at once, feels it all as one great blooming, buzzing confusion; and to the very end of life, our location of all things in one space is due to the fact that the original extents or bignesses of all the sensations which came to our notice at once coalesced together into one and the same space." Of course, that "one and the same space" is the brain, and in this chapter, we'll talk about how it learns to impose order on the "bignesses of all the sensations" of the outer world that it is born into.

Since no one *remembers* what it was like to be born into the "blooming, buzzing confusion," it's really hard to appreciate how

---

5. By comparison, chimpanzees are born with brains 36 percent of their adult size, and our more distant primate cousins, like macaques, are born with brains about 70 percent of their adult size.

much your experiences have shaped the way you understand the world around you. But if you've ever had a conversation with someone about an event you *both* participated in that left you feeling like one of you was delusional because your stories were so different, you might have a hint. This can be insanely frustrating because—let's face it—our own brains are *really convincing* when they construct *our personal version* of reality. Remember the Dress?[6] Though it can feel like gaslighting when someone has a different reality from yours, it's also entirely possible that you both were reporting your version of the truth.[7] At the end of the day, the way people *remember* a story reflects differences in the way they *experienced* the original event. The scientific explanation for this boils down to differences in *perspective.*

The word "perspective" seems particularly appropriate to describe the phenomenon that occurs when a brain becomes shaped by a particular set of life experiences, because it can refer both to the physical place a person occupies in space and to the mental place they interpret their experiences through. Other terms, like *point of view,* capture the same idea—that we may experience the same event from a different "place" that transcends the sensory and cognitive domains.

This shouldn't come as a huge surprise, based on what you've learned so far. In the first half of the book, we discussed how different brain designs shape the way we come to understand the world and behave in it. Then, in the previous chapter, we talked about how differences in the way brains focus influence which pieces of information will capture a person's attention and thus drive their conscious experience of an event. Now we're going to learn how our life experi-

---

6. As I promised, in this chapter, we'll talk about why people see the Dress differently.

7. By no means do I want to insinuate that people don't lie. Some do. I'm only trying to make the point that there are also other—honest—processes that can leave someone with a different memory of an event.

ences shape our brains, creating the lens through which we see the world, literally and metaphorically.

Before we get into the details about how this works, let me take a step back and remind you of the interpretive processes we discussed in "Introductions." Before you knew much about how your brain worked, I painted a picture of a powerful but discrete and finite information-processing machine that tried its best to understand a continuous and infinite world by taking a series of low-resolution "snapshots" of the physical world and connecting the dots. In this chapter, we'll talk in more detail about how our experiences inform this dot-connecting process. But please keep in mind that when I talk about "interpretation" in your brain, I don't mean the kind of deliberative interpretation that leads you to conclude things like "This person *said* X, but I know what they really *meant* was Y."[8] I'm talking about a much more *pervasive* type of reality construction—the kind that happens when every single packet of information from your sensory neurons arrives in your conscious awareness. At the end of the day, my goal is to help you remember that your brain isn't just passively viewing the world . . .

It is *creating* your understanding of reality through a lens shaped by your life experiences.

Most of the ways this happens are so fast, and so automatic, that you don't know which part of what you *perceive* has actually been detected in the world "out there" and which part your brain has interpreted. For a low-tech and nonthreatening example of this, take a look at the picture on the next page.

What do you see?

Most of you will describe something that looks like a three-dimensional cube, made up of black lines, floating in front of a black

---

8. This certainly can happen, and we'll talk about why when we discuss the challenges of communicating with others in the final chapter, "Connect."

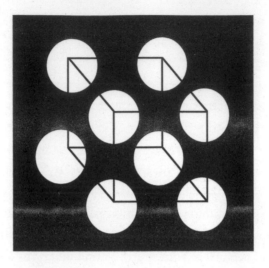

background with white polka dots. But if you focus on the place where one of the edges of that cube travels between polka dots, you'll see that the lines that make up its edges don't actually exist on the page. Instead, what is printed is a bunch of white circles interrupted by black line segments, which makes them look like poorly sliced pizzas. The box is something your brain *constructs* based on what it expects to see.

This illusion and the dozens of others you can find on the Internet are some of the fun examples, created by psychologists who study visual perception, that showcase the fact that your brain makes stuff up all the time. You *perceive* depth and motion in the static, two-dimensional images that flicker across your television, just as you seamlessly fill in the blanks during a cell-phone conversation in which much of the information from the speech signal gets clipped out. These are all shortcuts that our brains generate to make sense of the incomplete data they take in. But as you'll read in this chapter, there can be serious costs to these shortcuts, particularly when a brain is trying to operate in an environment it hasn't adapted to. In the next section, we'll discuss where these shortcuts come from by describing the ways each brain learns what to expect based on its previous experiences.

How you learn

Most people I talk to have a clear opinion about what *type* of learner they are. The thing is, when they describe themselves using labels like "visual" or "tactile" learners, what they're really referring to is their preferred modality of *instruction*. From the classroom to YouTube videos, most people associate *learning* with the kinds of explicit, instruction-based formats that having language enables humans to use. But the vast majority of our learning is much more passive than these activities. In fact, the most *expert* learners of all—babies—can't even follow instructions because they don't have a fully functioning language system on board.

From a neuroscience perspective, learning is *any* process by which your experiences change the way you will think, feel, or behave in the future. And if you look closely enough at brains at work, you'll find evidence of learning—and forgetting—in every waking moment of your life.[9] This is because every experience leaves its mark. Much like a walk down the beach leaves a record by pushing millions of grains of sand into slightly new locations, each of your mental experiences creates *physical* transformations in the connections between your neurons, which influences their subsequent communication.[10]

One of the most essential ways your experiences shape your brain is through a process called *Hebbian learning*. In essence, Hebbian

---

9. A lot of important learning and forgetting also happens when we sleep, but since much less is known about individual differences in those mechanisms, I've chosen to omit them to save space.

10. I borrowed this metaphor from Andrea, who uses it when he teaches, and I wanted to make sure he gets appropriate credit.

learning is the biological mechanism that allows your brain to keep a running set of statistics about how frequently things occur in your environment. Much like sports teams keep statistics of their players and use them to make decisions about who to start and who to trade, your brain has a way of "counting" the frequency of occurrences of different types of events and using this system to figure out what's most likely to be happening, given the incomplete information it receives.

Fortunately, your brain's way of taking statistics doesn't require any counting on your part. Instead, the work happens in the connections between the gossipy neurons—in the spaces that determine who is talking to whom, and how loudly. As you might recall from "In Sync," timing is really important for organizing such communication. As it turns out, it's also really important for learning. When two neurons in close proximity become excited at approximately the same time, the connections between them will strengthen, increasing the likelihood that the message of one will be picked up by the other. Though the actual principles of Hebbian learning are a bit more nuanced than this, I always remembered the catchy slogan I first learned as an undergrad: "Neurons that fire together, wire together." And the more often this happens, the stronger the connection between the two neurons will grow. This is your brain's way of connecting the dots. It assumes that if events A and B virtually always occur at the same time, they are part of the same "neural event." Once this happens, even if your brain only gets evidence that A is going on in the outside world, it is likely to assume that B happened as well, and will *create* that experience for you, much the same way the participants in Cecere's experiment *saw* two flashes of light when two tones were presented in different alpha windows, even if there was only one light presented.

As Hebbian learning continues to influence your brain over your life span, the strength of the billions of connections in it comes to function as a *massive* database that reflects the likelihood of everything you've experienced in your lifetime. My brain, for instance, needs very little evidence to understand what's happening when I see someone walking a dog in my neighborhood. Because I walk my dogs almost every day, and often see others doing the same, there are *thousands* of instances of such events in my database. The result is that the network of neurons dedicated to identifying dog-related activities in my brain is very well connected. It helps me with the visually demanding task of understanding what kind of complex, three-dimensional critter-in-motion is attached to the string a person is holding, despite the variety of sizes, shapes, and colors that dogs come in.

To be honest, I took this important (to me) adaptation for granted until the time I saw a guy in my neighborhood walking down the street with two *goats*! Because my brain had already filled in the blank for "what's likely to be at the end of that leash," I had to stop and stare with my "does not compute" face for a solid second before I could figure out what the *actual* hell I was witnessing. And it's not that a goat is harder to recognize than a dog.[11] If I were driving through the country and saw a barn with a grassy pasture in front, it would probably be much easier for me to recognize a goat in the field than it would be to understand that I was seeing a Bernedoodle. Because after forty-plus years of life as an animal lover, my brain has acquired quite a lot of information about when and where one is most likely to encounter different types of animal friends. And at the intersection of the search including "my neighborhood" and "walks on a leash" now lie two answers: "DOGS" in all caps, because it is thousands of times more likely than the other exciting possibility, "goats."

These shortcuts our brains make are critical to our survival. Even

---

11. There are about three hundred different breeds of goats, but they do not vary as much in size or features as dog breeds do!

if we *could* sample all of the bits of energy in the world around us and build an accurate representation based on the bottom-up, tree-level details, doing so would take *so much time* that the world would have changed by the time we understood it. This would leave us making decisions based on the world as it existed moments ago. Crossing the street, with or without a dog, could be deadly.

But much like a specialized brain region loses the ability to be "multipurpose," as an experienced brain becomes adapted to a specific environment, it can lose its ability to understand things it isn't regularly exposed to. The research of my brilliant and inspirational colleague Patricia Kuhl has shown that this is precisely what happens when infants are immersed in their native languages. All infants are born "citizens of the world," she says, because they can *hear* and discriminate between the different speech sounds that occur in *all* of the world's languages. But they're not very good at any of them to start out with. Then, as they gain experience listening to one particular language,[12] their brains adapt to the sounds in their native tongue. But as quickly as this happens, they begin losing the ability to hear or produce the sounds that they are not exposed to. By the time babies are six months old, you can already see signs that their brains are becoming *fine-tuned* for the sounds in their own environments. And though older children, and even adults, *can* learn the sounds of new languages later in life, it is much more difficult to do so. For this reason, most who do will have accents that persist—and the fact that they can't even *hear* their own accents makes it all the more difficult to change the way they speak.

Now that you understand something about the way Hebbian learning changes the shape of your brain based on your experiences, and what the costs and benefits are of such adaptation, in the next

---

12. Don't worry, we're going to talk more about the bilinguals who make up more than half of the world in just a second!

section we'll talk a bit more about the types of experience your brain adapts to.

What types of experience shape your brain?

Before we delve further into how our unique constellations of experiences influence the way we come to see the world, I want to be specific about what "counts" as an experience when it comes to developing your perspective. Put simply, you learn from all of your *neural experiences*. From your brain's perspective, it doesn't matter whether the signals passing through it originate from something you've seen in the outer world or whether it's just a bus-ride-induced fantasy. Each of the corresponding electrical storms shapes the landscape of your brain's database.

If you think about it, you can probably intuit why this is the case. When you remember an embarrassing or painful event, for instance, you might re-experience some of the emotions associated with the original event.[13] It might even make you blush or tear up! This is because the process of *retrieving* a stored memory places your brain into a state that strongly resembles the one it was originally in when the memory was recorded. Your brain counts this re-experiencing of the memory as a second learning event. Much like walking down the same path on a beach for the second time would both blur the details of the original trek and make the path more discernible, this re-experiencing of the original memory both changes its nature and increases the likelihood that the event will be retrievable in the future. Through this mechanism, memories, and even completely

---

13. Once I met Jeff Bezos at a scientific event and I was *so uncool*. "My husband really likes his Kindle," I said in the middle of a much more exciting conversation about rockets. Oof.

imagined events, can create learning effects that are similar to those created when your brain processes information in real life.[14]

I was able to use this knowledge *once* to achieve a pretty big parenting victory. When Jasmine was about four years old, she started taking gymnastics classes. She was incredibly graceful and strong, but she was also nearly twice as big as most of the kids her age. This made some of the strength moves really hard, and one in particular—a pullover—was preventing her from moving up to the next level.

For those of you who managed to dodge gymnastics, let me try to describe, with words, this very visceral process. Doing a pullover involves a freakish amount of core and upper body strength, and a bar. It starts with your feet on the ground, and your hands on the bar. Then, in one fell swoop, you simultaneously pull your chest toward the bar while jackknifing your feet into the air and pulling yourself, *feet first*, backward over the bar.[15] Jasmine had mastered all of the other skills at her level but couldn't move up without being able to do a pullover. She practiced for months—at the playground, during recess, and every other chance she could get. She would get *really close*, but she couldn't quite get her butt up over the bar without a little help from the ground. Then, on the night before the last class of her session, she confessed that she was feeling bummed out that her friends were going to move up without her because she couldn't do a pullover.

I'll admit it. I freaked out a little bit—but only on the inside. Jasmine was raised by a single mom in graduate school. I don't think either of us was particularly "failure deprived," but man, did I hate to

---

14. The really good news here is that if I ever do see Jason Momoa with beach drinks in real life, my brain will be able to recognize what's happening *right away*. The bad news is that I will almost certainly do something embarrassing very shortly thereafter.

15. I'm almost positive I'm doing a terrible job explaining this. If I've left you wondering, you can search "How to do a pullover" on YouTube—but don't be fooled by how easy they make it look!

see that little face feeling disappointed. I do believe that *sometimes* hard things can be achieved with hard work, but as a person with almost no physical talents, I also understand that there are some hard limits on that idea.[16] In my moment of desperation, I latched on to something I had learned about mental imagery and athletics in one of my graduate classes.

"Can you imagine what it would feel like to do a pullover?" I asked her as I tucked her into bed. "Yes," she said. "Well, then you can practice in your mind!" I told her. We walked through the process together. I asked her to remember what it felt like when she had gotten over the bar with a little help from her teacher. Together we painted the mental picture—kick and pull, tuck and *swoosh*.

To be perfectly honest, I *never* really believed it would work. I'm just one of those people who has a hard time doing nothing. But I'll be *damned* if she didn't do a pullover in her next class, and pretty much every time she tried thereafter.[17]

The memory of the pullover that we imagined into being has since become a well-worn path in both Jasmine's and my brains. It's one we are quick to remind each other of when we find ourselves worrying about something—which could be considered mentally practicing for things you *don't want* to imagine into being. As you continue to learn about the types of mental experiences that shape the way you come to understand the world, try to remember that as far as *your brain* is concerned, the realities you remember, imagine, and fret about all "count" as the data your brain uses to adapt to your

---

16. Unfortunately, there are also some pretty serious institutional barriers that get in the way in different circumstances.

17. Of course, there is no "control" experiment in this anecdote. It's possible that after so much practice in real life and a good night's sleep, she would have gotten it anyway. And no degree of mental practice can override the physical limitations of the body. I can't fly in real life, no matter how many times I practice in my dreams. There *is* a significant amount of experimental research, however, that does show improvements after mental practice.

environment. In the next section, we'll discuss more about one particular type of experience—language—and the profound effect it can have in shaping your brain.

Assessing your language experience

The central point of this chapter is that our life experiences shape our brains in ways that make them better suited for dealing with similar experiences in the future. Unfortunately, it would be impossible for me to create an assessment that would capture *all* of the different life experiences that have shaped your perspective. Even if I could, the vast majority of them would be things that haven't been well studied in the lab. So I've decided to zoom in on one experience that almost all human beings share, which is known to have a *pervasive* influence on your mind and brain—the language or languages we speak. This is because language is so central to the way we think, feel, and behave that we spend the majority of our waking hours using it. Below I've selected a few items from the Language Experience and Proficiency Questionnaire (LEAP-Q) we use in the lab, which was developed by Blumenfeld and Kaushanskaya.

ABBREVIATED LANGUAGE QUESTIONNAIRE

1. List all of the languages you know, in the order that you learned them.
2. Thinking of an average week, what percentage of the time do you spend using each language? This includes listening to music or watching television in the language, not only speaking it. (Sanity check: The percentages across languages should add up to 100).

3. If you speak more than one language, at what age did you first learn to speak in your second language? Repeat this question for the third language and beyond if you know more.

4. If you are conversationally fluent in more than one language, at what age did you first become fluent in your second language? Repeat this question for the third language and beyond if you are conversationally fluent in more than two languages.

5. If you speak more than one language, rate your proficiency in your second language on a scale of 0 to 10 for the following skills:
     a. speaking
     b. listening
     c. reading

(Note—If you don't have any experience in a second language, put zeros here.)

How diverse are your language statistics?

It doesn't really make sense to talk about how "typical" you are in this space, because to do so, I would need to know where you're coming from. If you live in Luxembourg, for instance, and you reported only knowing one language, you would be in the vast minority. In many American cities, this would not be the case.[18] And this makes sense

---

18. U.S. Census data on language use is pretty limited. They only ask whether there is a non-English language spoken in the home, and then ask participants to choose which of four categories describes how well they speak English: "not at all"; "not well"; "well"; and "very well." There's no information about how well

if you consider the fact that this chapter is about adapting to *your* environment. Remember, language is just one example of this adaptation. But since it relies on a set of statistics that we use constantly throughout the waking hours, it can provide a good model of how adapted your brain is.

If you do not know more than one language, or if your second language knowledge is limited (for example, your proficiency scores are less than 4), or acquired later in life (for example, post-adolescence), your brain is more narrowly tuned to your first language than if you have more diverse language experiences. One benefit of this is that your brain is likely *better* prepared to use that single language than the brain of someone who learned more than one language would be. Roughly speaking, the reason for this is that people who speak multiple languages have more options to consider when using their statistics to comprehend or produce one language.[19] They need to resolve competition between them before using any specific language. This means that it takes them a fraction of a second longer to access any piece of linguistic information they need to use, even in their most proficient language.

But as you'll learn in this chapter, there are also benefits to being broadly exposed to different types of statistics. Not only do people with exposure to multiple languages have a richer set of behaviors to choose from,[20] they are also likely to consider *more* information when deciding how to behave—such as which of their languages they think is most appropriate in the current context. But the cost of this

---

they speak the other, non-English language, or how frequently they use either language.

19. This is further complicated by the fact that cross-linguistic interactions create competition when the bilingual brain tries to select information in one language or the other.

20. I'm assuming that the benefit of being able to speak a language, and the opportunity it opens up for being able to communicate with someone in that language, speaks for itself.

consideration of diverse ways of responding "in the wild" can certainly add up. In short, having a brain that is more widely exposed may slow down processing in any particular environment or context, but it also allows a person to be prepared for a larger number of situations.

So how narrowly or widely "tuned" is your brain to language? In considering your profile of responses, one thing I'd like to point out is that—like most other things we've discussed in this book— language use is a multifaceted concept that can't be well characterized by the single axis that "monolingual" versus "bilingual" or "multilingual" seems to reflect. In our lab, we have explored differences in brain and cognition that are related to four aspects of language experience: (1) How early—if ever—a person was exposed to a second language; (2) How proficient a person is in their second, or least-dominant language; (3) How frequently a person currently uses each of the languages they know; and (4) How similar the languages they speak are. Each of these aspects relates to the way your brain uses your previous experiences to shape the way you understand the world and operate in it.

But notice that none of these questions is about how *many* languages a person speaks. The reason for this is that the way different language experiences shape your brain depends on the other factors. In the remainder of this chapter, we'll use language as a model experience to discuss *why* these factors matter and provide insights about a broader range of life experiences that your brain adapts to.

Why age matters: The differential impacts of early and late experiences

Language provides an interesting model for exploring how our experiences shape our brains, because—for most of us—it will continue to do so throughout our lifetimes. For example, most English-speaking adults know the meanings of between 20,000 and 35,000 English words,[21] yet there are more than 170,000 words *in current use* in the English language. This means that if you are a person who likes to read, or listen to podcasts, or even engage in conversations on subjects you are unfamiliar with, you are very likely to continue bumping into words you've never heard before. Maybe you even learned a few in this book? Yet most of us understand that it's much easier to learn languages when we are children than it is to pick one up as an adult. This raises the question: How much learning happens in the earlier years of life, when we're trying to make sense of the "blooming, buzzing confusion," and how much can we adapt to later?

The short answer is that there are different *windows* of adaptation in different parts of the brain. To simplify, we can sort brain regions into three types, based on how much, and for how long, they are open to experience. The first type, made up almost entirely of the parts of the brain that regulate the functions that keep you alive, are *experience-independent*. These are the parts of the brain that regulate your critical functions like breathing, heart rate, and body temperature, which do not vary much across different environments.

Next we have the *experience-expectant* regions. These are the parts of the brain that are predestined[22] to learn to interpret specific types of information about the world "out there," because they are hardwired to receive information from our senses. For instance, in typically developing babies, light coming in through the eyes is carried to the occipital cortex at the back of the brain, sounds coming in through the ears are carried to the auditory cortex in the temporal

---

21. You can get a quick estimate of the size of your vocabulary by visiting testyourvocab.com. Do you know what "terpsichorean" or "tatterdemalion" mean? I did not.

22. Unless you are a ferret and some curious neuroscientist decides to rewire you . . .

lobes on the sides of the brain, and smells coming in through the nose are processed by the olfactory bulb, located at the bottom of the front of the brain. The fact that we have to *learn* to recognize the things we see, hear, and smell allows human babies to develop expertise in the environments they were born into. And as the French documentary *Babies* depicts, there are both remarkable similarities and interesting differences in the environments of infants raised in different parts of the world.

However, because our brains evolved before things like airplanes and the Internet made it easy to transport ourselves from one part of the world to another, many of the experience-expectant regions also have "critical periods"[23] for receiving input. At the beginning of life, they are waiting for data and are incredibly malleable. But as you age and these areas amass information about the world around them, they become more and more entrenched in the processing of the kinds of things they expect to see and are less influenced by new experiences in the outside world. The strength of early experience on experience-expectant brain regions was convincingly demonstrated in a series of experiments in the 1970s, in which kittens were raised in very specific visual environments—like rooms with only vertical lines, or barrels with objects on the walls that rotated only to the left. Under these conditions, the brains of the kittens became so tuned in to the restricted kinds of stimuli they were exposed to, they couldn't *see* the things they hadn't been exposed to earlier in life—like horizontal lines, or objects that were moving rightward! Thankfully, we don't raise babies in barrels, but Patricia Kuhl's research showed a similar effect based on the speech sounds that babies are exposed to. As a reminder, after the first year of life, human infants lose their

---

23. Many scientists prefer to call these "sensitive" periods, as it seems to be a matter of degree of being open to experience rather than an all-or-none phenomenon. Also, individual brains differ in when, and to what degree, they stop being open to input.

sensitivity to speech sounds that don't occur in their native languages.

Fortunately there *are* parts of our brain that remain malleable over much of our life span.[24] These are the *experience-dependent* parts of the brain. Among them are most of our cortical "association" areas, including those that allow us to acquire new vocabulary words throughout our life. One of the most critical *experience-dependent* regions is the frontal lobe, which—as you learned in the previous chapter—supports much of the flexible behavior that characterizes human adaptability. And as you might guess, the basal ganglia nuclei are also experience-dependent. In fact, they are arguably among the most adaptable brain regions because they are rich in the dopamine communication signals that increase neural plasticity. In the next chapter, "Navigate," you'll learn about how important this is for shaping your brain's decision-making processes. Meanwhile, we're going to discuss the important work your brain needs to do *before* you can make an informed decision about how to behave. Because before your brain can learn what to do next, it's got to have a good idea of what's going on right now. In the next section, we'll return to our discussion of shortcuts and learn about some of the ways our experiences shape the way we learn to understand the world in the visual domain.

Developing perspective: How your environment shapes what you see

For a surprising—but fun—example of how our experiences shape the way we understand the world, let's go back to the Dress. As it

---

24. This is not to say that our brains are *equally* malleable across our life span, or that our early and late experiences will have equal effects on the way we behave. But it is true that certain parts of our brain remain more open to experience than others.

turns out, at least *part* of the reason some people see the Dress as white and gold, while others see it as blue and black, can be traced back to differences in our experiences. In fact, the whole process of seeing color likely involves much more interpretation than you think. This might surprise you if you learned that the colors we see correspond to specific wavelengths of light. You may even have learned that typical color vision includes three different types of receptors (or cones) located in the back of the eye that respond preferentially to long, medium, or short wavelengths of light. This seems like a straightforward way for figuring out what color something is without connecting any dots. How can the *color* of something be open for interpretation?

Fortunately, the way we understand what color something is, is not that simple. If it were, we all *might* be able to agree on the color of the Dress, but we would also agree that a green apple turns red at sunset and bluish in a shadow.[25] Because the fact is, as the nature of the light that bounces off an object changes, so do the wavelengths that reach our brains through our eyes. Thankfully, one of the things our brains learn from experience is that the properties of the light bouncing off an object are more likely to change than the color of an object itself. And so, to adjust for different lighting situations, it uses a *shortcut*: It takes a survey of all of the wavelengths in a particular context and uses the differences between them—rather than their absolute value—to figure out what color something might be.

What makes the picture of the Dress tricky for your brain to interpret is that there isn't a lot of context in the picture to gauge what kind of lighting is bouncing off of it. Under this circumstance, people's brains make *different* assumptions, automatically, about what

---

25. For the sake of time, I'll avoid the intricacies of how different wavelengths of light are absorbed by the Earth's atmosphere when the sun is low on the horizon, and so forth. Suffice to say that under different conditions, the characteristics of the light that bounces off an object and into your eyeballs can change quite a lot.

the light is like in the picture. Those of you who see a white-and-gold dress do so because your brain assumes, based on *your* lifetime of experiences with light sources, that light is coming from behind, and that the dress is in a shadow. To "fix" this, it automatically subtracts out the dark blue and black hues and leaves you seeing white and gold. Others, like me, who see the Dress as blue and black, assume that it is well lit from the front or top, possibly through some source of artificial lighting, and so we make no such subtraction.

So what kind of life experiences might shape our assumptions about lighting? Writer and vision researcher Pascal Wallisch tested one interesting hypothesis by exploring whether people who tend to wake up early in the morning (the "larks") would see the Dress differently from those who wake up late and stay up late (the "owls"). His assumption was that larks would have more experience with natural lighting, and thus would be more likely to assume the dress was in a shadow and see it as white and gold. On the other hand, owls, who spend more time awake after dark, would have more experience with artificial lighting and would be more likely to see the Dress as black and blue. To test his hypothesis, he asked 13,000 people what color they thought the Dress was, and what their normal sleeping habits were. When he did, he found a small but reliable effect that was consistent with his intuition. The larks were more likely to see the Dress as white and gold, while owls were more likely to see it as blue and black![26]

It's also worth noting that much in the same way you can't *stop* seeing the black cube at the beginning of the chapter just because I tell you it's not real, you are unlikely to be able to "flip" the way you

---

26. Note that I am an extreme lark and I see blue and black, and Wallach admits that he is an extreme owl who first saw the Dress as white and gold—however, as he notes, given a lifetime of visual experiences, it is reasonable to expect that only a small amount of variance can be explained by your sleeping and waking preferences. For example, many larks also spend a lot of time working indoors, and many owls are forced to wake up earlier than they would like to work nine-to-five jobs.

see the Dress just because you learned about the lighting effect. This is a salient example of a situation in which goals or instructions can't override the automatic processes your brain has learned from experience. We'll talk more about this in "Navigate"—but if someone were *very* interested in changing the way they see the dress, they'd need to systematically feed their brain a lot of the kinds of lighting experiences (natural or overhead lighting) that are thought to shape these early perceptual processes—and they'd be working against a lifetime of learning and their critical periods for vision.

While I can't imagine why a person would be motivated enough to do so for the purposes of the Dress, I sincerely hope that some of you might be interested in modifying the way your brain comes to understand about more complex relationships, including the implicit biases we form in our *experience-dependent* brain regions around race, age, gender, and sexual orientation—to name a few. Even though these biases involve the way we learn to associate higher-level concepts that occur at the same time, or in the same contexts, with one another, they can still *influence* our early, perceptual understanding of the world in disturbing ways.

One glaring example of this, which has been *repeatedly* demonstrated in laboratories around the world, in different populations and under a variety of conditions, is that people are more likely to report *seeing* a weapon when an ambiguous object is presented next to a Black face (in space or time) than when it is presented next to a White face. The effect was first demonstrated by Keith Payne[27] in 2001. Across two experiments, Payne showed 60 non-Black participants a series of black-and-white pictures of tools or handguns, flashed quickly on the screen for one-fifth of a second, and asked them to indicate what they saw. The catch was that in both experiments, a

---

27. This name seems so appropriate, given the uncomfortable truth that his research uncovered.

picture of a Black or White male face was presented briefly before each to-be-recognized object. The research participants were told that the faces were just a cue that the object was coming; they were not expected to relate to the objects in any way, and in fact they didn't. Black and White faces were presented equally often before both tools and handguns. Despite this, Payne's participants found it significantly *easier*, based on their reaction times, to recognize a handgun when it was presented after a Black face than after a White face. The gun was also more easily recognized than the tool following the Black face, though both were equally easy to perceive when they followed the White face.

Though the size of this effect was pretty small,[28] what it reflects about learning and the brains of the participants is *largely important*. The fact that guns shown after Black faces were the *easiest* thing in the experiment to recognize suggests that, on average, the neural databases of the participants contained a strong enough link between Black faces (A) and guns (B) that a shortcut is created in their brains. In other words, the most straightforward explanation for *why* people were faster to recognize a gun following a Black face is that when they saw the Black face in isolation, their brains had already started to fill in the blank and construct the concept of a weapon.

The chilling, real-world implications of this "shortcut" were made even more salient in the second experiment, which forced participants to make faster decisions about whether or not an image presented to them was a gun. Following a Black face, participants mistakenly identified a tool as a gun 37 percent of the time (more than 1 in 3 times), whereas the reverse mistake, identifying a gun as a tool, happened 25 percent of the time.[29]

---

28. Presentation of the Black face before the gun decreased recognition time by about 20 to 30 milliseconds.

29. Following White faces, participants made the same number of errors identifying tools or weapons.

The deadly consequences of this are obvious to anyone with access to the news.[30] And unfortunately, one critical question that is largely *unanswered* by this original research is: How do we fix it? One place to start is to figure out *where* the data that drives these biases comes from. Even though plenty of Americans own guns, it's difficult to believe that the average college kids in these studies had many (or any) real-life experiences with Black men and guns. So, where do these shortcuts come from?

To answer that question, let's return to our notion of what "counts" as an experience. To put it simply, the less experience you have in real life with a particular type of person, place, or thing, the more likely it is that your brain's database entry for that topic is based on what you see on television, or read about in the news, on social media, or in *fictitious* depictions. Remember that your brain doesn't really care if you're experiencing something, remembering it, or imagining it—all of these mental experiences count. And so, if the Black faces you see on television are more likely to be holding a gun than a stethoscope,[31] your brain will assume that this is true of the world and incorporate it into the experience-based lenses through which you see.

In this way, many of our brains *literally* become shaped by the systemic biases of our society as we consume the versions of reality created by others.[32] And these biases can influence the way we understand the world in ways that are as fast and automatic as those that interpret the color of the Dress. Which brings me to another important distinction—the people who participated in this research and have these types of shortcuts in their brains do not *necessarily* hold conscious, explicit ideas about what kinds of people carry guns. In

---

30. Malcolm Gladwell covers this to some extent in his book *Blink*.

31. Enormous thanks to Shonda Rhimes for helping to correct this stereotype (in my brain, for sure!).

32. Notably, these others tend to be privileged White males creating media from their point of view.

fact, your explicit beliefs and your experiential database are perfectly capable of contradicting one another—an idea we'll return to in "Navigate." This makes it unclear whether, or how, the trainings on implicit bias that are now delivered in courts and in the workplace might have any effect on the shortcuts our brains have formed.[33] Much like telling you *why* you see a white-and-gold dress even though it was actually black and blue doesn't change what you see, becoming aware of your implicit biases is unlikely to change the way they automatically influence you. The best one can hope for from such an education is that it will change your *awareness*, and by virtue of this, make you think twice about your behavior. We'll talk more about the relation between knowing better and doing better in the next chapter. For now, let's wrap up what we've discussed in this one by looping back to our conversation about language experiences.

Summary: Narrowly tuned brains are well prepared for specific environments, while widely exposed brains consider more options

You might think of the implicit biases we discussed in this chapter as the result of an over-adaptation that occurs when a brain becomes entrenched in an environment that is sufficiently narrower than the one you would like to operate in. This is kind of like the specialization process we discussed in "Lopsided" when a brain region becomes better and better prepared to execute fewer and fewer tasks. The difference is that in this case, it's giant swathes of the brain that are being prepared to exist in a particular place and time. And based on what we've discussed about how fundamentally experiences

---

33. It is worth noting that the efficacy of such programs *is* being investigated, but the results to date do not provide a clear mechanistic explanation for how such trainings might work.

shape your understanding of the world, I believe that we'll need to do more than just read books[34] to correct these shortcuts in our brains. For one thing, we need to be more intentional about what kinds of experiences feed our brains. What are the underlying associations in the ideas we consume?

Another way to expand our databases is to expose ourselves to diverse, real-world experiences and allow narratives told from different points of view to shape us. Returning to our model example of language experience, we might draw some inferences about what brains with more diverse life experiences would look like. The fact is that people who regularly speak more than one language need to learn at least *two* sets of linguistic statistics. And doing so is not without a cost. There's considerable evidence that if you measure proficiency in one language in a child who learns two at the same time, their developmental trajectory is a bit slower than those who are learning only one. And even in *highly proficient* bilingual adults, it can be a bit more difficult to access either language than it is in a monolingual brain.

This is because the statistics for a bilingual's two languages are not kept in nice, separate packages in their brain. In fact, they are so intimately intertwined, hearing or thinking about any word in one language automatically activates the associated neurons in their other language to some degree. And when a bilingual wants to speak in the language they are less fluent in, they need to override stronger, and more automatic, activation from their dominant language. In other words, when a bilingual speaks in their non-dominant language, it's a little bit like having to name the color a word is printed in, when the word and the color it's printed in don't match!

Bear with me, I realize I'm not making this sound very appealing.

---

34. Don't get me wrong—I'm a firm believer in a good nonfiction book as a way of educating yourself. I just don't think it can stop there if you want to change, for reasons we'll discuss in "Navigate."

Andrea and I have studied basal ganglia signal routing in bilin-guals, and we have collected evidence that suggests that bilingual language experience can train a person's brain to have a stronger "rider." In fact, we showed that this improved signal routing in bilin-gual individuals made them quicker than monolinguals when exe-cuting new *mathematical* tasks. The fact that people with diverse language experience have multiple ways of expressing themselves may create more conflict in their brains—but it also trains them to be more present to the *context* they are operating in. Bilinguals can't just get by with the automatic spreading of activation from A to B and still use their less-dominant language with any precision.

In a recent study I published with my postdoctoral trainee Kinsey Bice[35] in 2020, we looked for evidence of such increased control in 197 brains—91 that belonged to people who knew only one lan-guage, and 106 that belonged to people who had experience with more than one language—based on their task-free patterns of neural orchestration. Overall, people with more diverse language experi-ences had higher alpha power, suggesting that they had greater amounts of brain synchronization coming from those inner-world control frequencies. And keep in mind, this is while their brains were *resting.* To add to what we've learned in this chapter about how brains "adapt," the influence of bilingual language experience on alpha power was highest in those who use both languages on a regular ba-sis. This suggests that we need not only to have a diverse database of experiences but also to practice *using* these experiences to operate in different contexts.

Although research on the broader effects of different language experiences on people's characteristic patterns of thinking, feeling, and behaving has been somewhat controversial in the field, we have argued that this is partially due to the fact that researchers don't

---

35. Who speaks four languages to varying extents—and two fluently!

always consider the different types of experiences that fall under the umbrella term "bilingualism." There does seem to be mounting evidence that people who use more than one language on a regular basis have brains that look and work differently from those who use only one.

While I'm not naïve enough to think that learning to speak multiple languages would correct the problems created by implicit biases, I do think it's important to become aware of the sources of information we're feeding our brains. Because it's entirely plausible that multilinguals provide a good model of what brains exposed to diverse sets of statistics look like. It may make us a bit slower to respond in any specific situation, but it may also force our brains to become more flexible and sensitive to the context they're operating in. In the next chapter, we'll discuss the implications of this more deeply, as we delve into the mechanisms your brain uses to make the big and small decisions that drive you through life.

CHAPTER 6

NAVIGATE

How Knowledge Creates Road Maps and Why We Don't
Always Use Them to Guide Our Decisions

Now that you're more than halfway through this book, I'm going to *hope* that you're too far in to turn back, and so I will take the opportunity to start asking some of the harder questions about you and your brain. In the previous chapters, I have given you *a lot* to think about. Most of it has been related to scientific findings about how your brain works. Here and there, I've even thrown in some unsolicited advice. And based on what we discussed in "Adapt," I know these experiences will have some influence on the way your brain is connected. But the real question I have for you is: Do you think any of the things you've learned will actually *change* the way you think, feel, or behave?

Maya Angelou, one of the most extraordinary humans I have ever had the privilege to listen to,[1] once said, "Do the best you can until you know better. Then when you know better, do better."[2] And

---

1. To be clear, I never met Maya Angelou. But Jasmine and I did have front-row seats when she came to give a lecture at UC Davis during my graduate studies, and it *felt like* she was talking to me. The experience, like the woman herself, was phenomenal.

2. I'm not sure when Maya Angelou first said this, but I suspect it became

though I return to her words for inspiration, I also realize that for most of us, the relation between what you know and what you do is *not* that straightforward. And in this chapter, we're going to start digging into *why*.

To do this, we need to build on what you've learned in the last two chapters about how different brains focus and adapt to their environment. From there, a logical question that arises is: How do different brains *use* their understanding of the current situation, along with the knowledge they've acquired from their previous experiences, to navigate through their lives?

Subjectively, some experiences certainly *feel* like they're more influential on how we behave than others. But we've spent quite a bit of time now discussing how our brains construct our realities in ways that we're not even aware of. So how does it all come together to influence the way we behave?

To better understand how what you learn does or doesn't shape your decision-making processes, let's start with an imaginary scenario in which you are explicitly taught a piece of information that has clear implications for how you *should* change your behavior. Say, for the sake of argument, that your doctor tells you that your blood sugar is high. This is not good news, because it puts you at an increased risk for type-2 diabetes, for heart disease, and for stroke. But according to your doctor, the condition is reversible. If you implement the lifestyle changes they suggest—decreasing the amount of sugar and refined carbohydrates in your diet and increasing your exercise—there's a great chance that your blood sugar will return to healthy levels. And when it does, you'll have more energy and feel better![3]

---

something like a mantra she said often. I do remember Oprah talking about the influence it had when Angelou said it to her!

3. As a reminder, in case you didn't read the Preface, I am not *that* kind of doctor. This is just an illustration.

So what happens next?

"Awareness is the greatest agent for change," influential author Eckhart Tolle says, and knowing you have a reversible health problem almost surely increases the *likelihood* that you'll make healthy changes to your daily habits. But much like the cost of paying attention is higher for some brains than others, the ability to control your behaviors based on instructions about what you *should do* is also much easier for certain brains to achieve than for others. We talked about this a bit in "In Sync" as well. How effortful is it to use those low-frequency, inner-world goals to direct the chorus of "voices" coming from the outer world? At the very least, I can say with a considerable amount of confidence that although awareness *can be* a catalyst for change, it certainly does not *guarantee* change. As Daniel Kahneman said of his highly influential book *Thinking, Fast and Slow*, "It's not a case of: 'Read this book and you'll think differently.' I've written this book, and I don't think differently." If the world worked like this, reading *my* book might even give you the information you need to accept yourself just the way you are.

Instead, in the real world, we sometimes behave in ways that are totally inconsistent with how we *think* we want to behave. And when we reflect on these moments after the fact, the truth about the relation between knowing and doing seems closer to what my childhood guru, cartoon hero G.I. Joe, taught me: "Knowing is half the battle," he would say after the moral lesson of each cartoon episode was revealed. But even as an eight-year-old fan of the show, I wondered what the *other* half of the battle was.[4] In what remains of this chapter, I'll do my best to describe the *entire* battle, as I understand it.

---

4. I read this quote in a great piece of writing, Ariella Kristal and Laurie Santos's commentary "GI Joe Phenomena: Understanding the Limits of Metacognitive Awareness on Debiasing," written in 2021. Their thoughtful piece, and the name of their relevant theory, clearly show that I am not alone in wondering what the other half of the battle was!

To dig into this complex topic, let's return to the idea of the two types of "control" in your brain, which I previously introduced as the "horse" and the "rider," because the way you navigate through your life's decisions is undoubtedly made up of some percentage of the horse's decision-making and some percentage of the rider's more controlled driving. In this chapter, we'll discuss the way your experiences shape the decisions that both horse and rider make that move you around in the world.

As you probably remember from the previous chapter, the horse is a metaphor I'm using for your more intuitive control system. The type of knowledge your horse uses to move through the world, called *procedural memory*, is responsible for the most fundamental and frequent types of navigating you engage in. Ironically, if you're trying to write about it in a book, procedural memory might best be defined as the types of things *you know* that are hard to describe verbally. For now, take it as a given that your horse has a lot of knowledge that helps you to navigate, but it can't tell you what it knows with words. From the muscle memory that shows you where to put your foot when you're walking on different types of terrain (or in heels) to the gut feeling that can drive your big and small decisions, your procedural memory helps you navigate through life in automatic and intuitive ways.[5]

When it comes to navigation, your horse's operation principles are pretty straightforward. It wants to *maximize* the chance that you win in the game of *The Brain Wants What It Wants*. This means driving

---

5. I've always marveled at the different ways people who *teach* procedural skills try to explain things, like how to ride a bike, or a horse, or how to execute a dance move. Ultimately, these things can't be learned through instruction alone. You need to *feel* them. Ironically, I'm bumping up against the same challenge when trying to describe the conscious experience of a procedural memory. If you've ever learned to ride a bike, or taught someone else, you might think of your procedural memory as all the critical bits of knowing that would be really hard to explain, like the timing between shifting your balance and turning the handlebars that results in a successful turn.

your decisions in ways that find the biggest rewards while avoiding pitfalls. If you are hungry, and there is a pint of delicious, rewarding, high-energy ice cream in the freezer, your horse will think eating that ice cream is an excellent idea. It's a horse—so things like diabetes or how you're going to feel in your jeans afterward don't factor into its decision space. In short, your horse and its motivation for rewards are certainly part of the other half of the battle.

In the first section of this chapter, we're going to talk about differences in how your horse-guided navigation system learns from your experiences. Because, as it turns out, not every horse learns the same way. In the next section, we'll discuss the parallel paths that drive the more intuitive navigation systems in the brain, and how some brains are more influenced by one or the other path.

Horse navigation: Learning by carrot or by stick

In "Adapt," we spent a lot of time discussing the early, automatic short-cuts individual brains take when they're trying to make sense of the blooming, buzzing confusion of the world "out there" based on their experiences. And in "Focus," we talked about how those experiences can also shape, and be shaped by, what we learn about the importance of various pieces of information. From your horse's perspective, these brain functions are a means to an end. To put it simply, your horse learns to pay attention to features of its environment that are associated with good or bad outcomes while at the same time learning to ignore things that have no effect on their decision space. Learning to distinguish the sounds of a language you don't speak, for instance, is unlikely to be useful. But learning the difference between "ba" and "pa" in native English-speakers might empower you to request a "banana"—a real and delicious food—instead of a "panana," which isn't a thing. But to learn which features of the environment are *important* to consider when deciding what to do next, your brain needs a way to link the environment it's in, and the choice it makes, to the *consequences* of that choice.

This process, called *reinforcement learning*, is one of the strongest influences on your horselike navigation systems. To move you through life in a way that maximizes your rewards, your brain engages in the following four-step procedure: First, it uses the database of your previous experiences to build the most accurate representation it can of the *important* features of the world around you. When bombarded with information on a busy street, for instance, doing a quick assessment of whether a person walking toward you is holding a tool or a weapon will motivate different actions on your part, while noticing what type of shoes they are wearing may be less consequential.[6] Second, based on your previous experiences in similar environments, your brain builds a representation of the repertoire of possible actions you might take. For example, you could smile and say hello to the person approaching you or engage in some other positive interaction. You might choose to remain neutral or pretend you don't see them. You could also avoid the person by walking (or running) in the other direction. Of course, there are any number of other actions available, from belting out "Halo" by Beyoncé to deciding to sit on the corner and sketch the scene. Whether your brain considers these actions has a lot to do with your previous experiences in similar contexts. Third, based on what you've learned from your previous actions,[7] your horse brain navigates the path it believes will bring you the most success by deciding which of the potential actions you can take is most likely to have a good outcome. Then fourth, your brain compares the actual outcome of this choice with the outcome it

---

6. The Italians in my family may never forgive me for suggesting that shoes are inconsequential . . .

7. We'll talk more about this mechanism shortly.

expected, and updates its database on the relative "goodness" of that choice accordingly.

Of course, many of the lessons we learned about how different brain designs influence the way we understand the world apply to this decision space. For example, whichever aspect(s) of the environment capture your attention will be weighed most heavily in your brain's decision about what to do next. And perhaps most critically for the discussion at hand, people with different life experiences will have different estimates of how good the outcomes of any of these actions may be.

For instance, if you are a bad singer, or even a good one who embarrasses easily, singing "Halo" in public may move to the bottom of the list. But when you don't have much experience with a particular situation, your brain starts to explore the possible actions through trial and error, to learn how good (or bad or ugly) the results may be.[8]

On a busy street, you might try out a smile. If the response is good, and not creepy, your brain will take note, increasing the likelihood that you will choose that option again in a similar circumstance in the future. But the kinds of notes it takes depends on the fourth and final stage of reinforcement learning. Because *very few* things in life are certain, and we rarely find ourselves in exactly the same situation twice, your brain learns by comparing how good what *actually* happened was to what you *expected* to happen based on your previous experiences. If the outcome is better than you expected it to be, your brain releases dopamine, the feel-good chemical. And as you may remember from "Mixology," that creates the learning signal that causes it to rewire, increasing the likelihood that you will choose that action again in a similar situation in the future. But if the result was worse than expected, your dopamine neurons would dip below their

---

8. This is the topic of the next chapter.

baseline firing rates, which would leave you feeling disappointed and weaken the connections to that action.

For example, if you chose to sing "Halo" in the middle of the street because you expected that you would sound just like Beyoncé, and that everyone would surround you and lift you in the air and hail you as their queen, the real-life outcome of that choice would probably put your brain into a tailspin. The result in your brain would probably make you feel more like you just got the "dance lesson" in the movie *Hitch*, when Kevin James starts showcasing his version of awesomeness and Will Smith looks him straight in the face and provides the dopamine-destroying feedback, "Don't . . . EVER . . . do that again."

As your brain fine-tunes its expectations about the outcomes of your actions, it builds something like a playbook with information about the best and worst actions you can take in any given context. And it uses this playbook, largely automatically, to guide your basic decisions about what to do when. From your first decision in the morning, like whether to press Snooze on your alarm to the words you choose to use when expressing yourself, your brain's four-step reinforcement-learning process makes it feel effortless to choose the most rewarding actions based on your previous experiences. And the process works wonderfully. In fact, reinforcement learning alone drives the decisions of most living, behaving creatures as well as most of the successful artificially intelligent systems we've built.[9]

But there's also something really crappy about living through the reinforcement-learning process that AIs don't have to deal with: If you are lucky enough to have a series of really good experiences associated with a particular choice, your brain learns to give that action a high expected reward value. And though this expectation will

---

9. AlphaGo, the remarkable computer program that beat the world's best Go player at his own game, is one of many AIs that was trained using reinforcement-learning algorithms.

make you *want* to make choices that lead to that experience again, the dopamine-based pleasure response you get each time is only based on the *difference* between your expectation and what actually happens. As a result, your actual experience of pleasure decreases the more often you get that really good thing. In other words, you only get the really big bursts of dopamine when something is *surprisingly* good, or if a known decision ends up being *even better* than you imagined it would be.

To put this a bit more concretely, no matter how much fun you *imagine* it would be to sing like Beyoncé, *being* Beyoncé would not be as fun as you might imagine. For one thing, your brain would have a whole different level of expectations. Not only does Beyoncé sing *exactly like* Beyoncé, she also wakes up looking like Beyoncé, and has a record of producing some of the most badass slaying performances in the history of the world. And because I have to suspect that her brain's reward system works just like ours, it's holding out for a better-than-normal-Beyoncé-level performance to give her a big dopamine burst! This naturally sets her up for plenty of disappointing moments and makes me wonder—would you trade places with Beyoncé for a day?[10]

Before you answer, let me tell you about one more critical detail that might influence both *how* and *what* your brain would learn from the experience of being Beyoncé for the day. As it turns out, the human brain has a critical fork in its reinforcement-learning path that is not present in most AIs. This fork corresponds to *two real pathways* in the brain that dopamine learns through. The first path, which I'm going to call the "Choose" pathway, works pretty much as I've outlined above. When the Choose pathway gets a dopamine reward signal, it strengthens the connections between an environment and the

---

10. My answer is yes ... but only if I could bring my own brain with me so I could take home the memories ...

chosen action, making you more likely to pick that play in the future. But there is another path that works in parallel, which I'll call the "Avoid" pathway. This dopamine pathway is composed of the kinds of receptors that inhibit, or turn down the volume on, neurons. So when you choose an action that is *less* awesome than you hoped it would be, and your dopamine dips below its baseline levels, the Avoid pathway actively learns that this is *not* a good action and weakens the connections. Horse navigation systems in *all* human brains learn in both ways—by "carrot," strengthening the likelihood that we will move in ways that bring us closer to good options, and by "stick," weakening the chances that we will turn in directions that bring us to the less-good outcomes. Critically for how your brain navigates, different dopamine receptors dominate the "carrot" and "stick" learning pathways. As a result, the differences we discussed in "Mixology" that drive chemical language communication also determine the extent to which one pathway influences the horse-based navigation in any individual brain more than another, or whether they contribute equally to guide your choices. In the next section, we'll describe some of the tests that have been used to measure carrot-and-stick learning, before getting into some of the real-world implications of how these different types of learners navigate.

Assessment: Are you a chooser or an avoider?

The best way to figure out the relative strength of your Choose and Avoid learning pathways is to take the test developed by Michael Frank and his colleagues, called the Probabilistic Stimulus Selection Task. We'll call it the PSS for short. Unfortunately, because this test is about how you learn from the outcomes of your decisions, it requires *live feedback* in a way that can't be provided in a book. If you'd like to figure out whether your brain learns by carrot or by stick, go to the "Research" tab on my website *before* you read the next section about how the test works.

The PSS has two phases. In the first, participants learn to select the most rewarding of two novel actions. "Actions" are represented by presenting two unfamiliar objects, and participants are told to choose one or the other. Of course, because they don't know which one is better at first, they start with a random guess. Then, after each decision, they get feedback. Sometimes they see the message "CORRECT" in big green letters, letting them know they made a good choice. And other times they see "INCORRECT" in red letters, letting them know they took a metaphorical ride on their tricycle down the stairs. This might sound like a cheesy way to replicate the real-life rewards and setbacks we experience based on our choices, but bear with me. The goal of this paradigm is to measure whether people learn *more* about the choices that lead them to CORRECT feedback or from those that lead them to INCORRECT feedback.

The trick is, there's not one right answer. Just like in real life, making the same choice in the experiment twice doesn't *always* lead you to the same outcome. This is where the word "probabilistic" in the task name comes into play. Instead, your likelihood of getting "CORRECT" or "INCORRECT" feedback changes depending on the action you pick. During the first phase, participants are given three choices to make, based on three pairs of six different actions. In one choice, the worst action of the six, which generates an "INCORRECT" response 80 percent of the time, is paired with the best action, which generates "CORRECT" feedback 80 percent of the time. This is the easiest choice for people to learn to make, because the difference between the outcomes of these two actions is large. The second choice, which includes one action that generates a "CORRECT" response 70 percent of the time, and another one that generates an "INCORRECT" response 70 percent of the time, is slightly more difficult to learn. And the third choice, which consists of one action that is "CORRECT" 60 percent of the time and "INCOR-RECT" 40 percent of the time, and another that is "INCORRECT" 60 percent of the time and "CORRECT" 40 percent of the time, is *really* difficult. No matter which item you pick, you get "INCOR-RECT" responses *almost* half of the time. It frustrates the heck out of participants who are sensitive to negative feedback.

During the first, learning, phase of the experiment, participants are given these three choices over and over again, until they learn to reliably choose the object with the highest chance of being rewarding (or the lowest chance of being incorrect). Then, to figure out *how* participants learned to make their decisions and which reinforcement-learning paths were involved, the symbolic actions get scrambled into different pairs.

In the second, decision, phase of the PSS task, each action gets paired with every other one. As a result, sometimes people have to select between the two best actions (the ones that result in positive feedback 80 percent and 70 percent of the time), and other times they have to decide between the two worst options (the ones that are only rewarding 20 percent and 30 percent of the time). Along the way, they also see every other pair in between.

And here's where things get fascinating, in my opinion: Even though the mathematical difference between 70 percent and 80 per-cent reward probability is essentially equivalent to the mathematical difference between 20 percent and 30 percent reward probability, the choices people make show that what they learn about the best op-tions is *completely independent* of what they learn about the worst op-tions! In fact, around 12 percent of the people we run this test on in the lab seem to learn primarily by carrot, using the Choose dopa-mine pathway to develop expertise about the reward likelihood of the best options, while another 12 percent of them appear to learn much better by stick, using their Avoid dopamine pathways to learn how to steer away from all of the worst options.[11] The rest fall somewhere between, using a balance of Choose and Avoid pathways to guide their decision-making. In the following section, we'll discuss how this relates to decision-making more generally.

Understanding the real-world implications of carrot-and-stick learning mechanisms

To provide a concrete example of how carrot learners and stick learners might behave differently, I've provided one example of a series of puzzles we created in the lab based on the Raven's Advanced Progressive Matrices—a test that measures reasoning and problem-solving abilities. The goal of these puzzles is to figure out which of the four options presented is the best completion of the matrix of images provided. To find the answer, note the way the images change as you move from left to right, and from top to bottom, across the matrix.

Some laboratory versions of this task are timed, and others are not. But since I'm only giving you one problem to solve, I'd encourage you to take as much time as you think you need. Once you've selected the answer you think is best, move on to the next page to figure out what it might say about how your brain learns.

---

11. These numbers come from an exploration of data from 365 participants in which I characterized anyone who performed at least 33 percent better on Choose than Avoid accuracy as a carrot learner and anyone who performed at least 33 percent worse on Choose than Avoid accuracy as a stick learner.

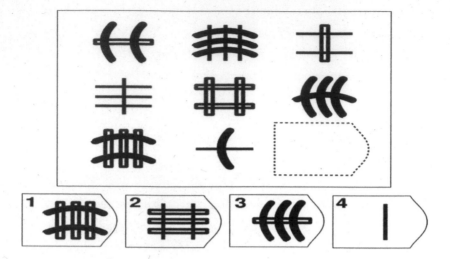

The correct answer to this problem is piece number 2. The reason this is the correct answer is that the pieces of this puzzle change according to the following combination of rules: As you move from left to right, the number of vertical markings increases—from 1, to 2, to 3, and then back to 1—while the horizontal markings decrease in number—from 3, to 2, to 1, and then back to 3. Even if you figured out only this rule, piece number 2 would be the only allowable solution to the problem. But there are other patterns to notice as well. As you move from left to right, the patterns of the vertical and horizontal markings also alternate in regular ways between hollow rectangles, regular lines, and bent, macaroni-looking shapes. Now notice how the number and the patterns of the horizontal and vertical lines also change in regular ways as you move from top to bottom.

Though there are a number of different ways you might happen upon the correct answer to this puzzle, research Andrea and I conducted, along with Lauren Graham, one of our former trainees, has shown that the stronger people's Avoid learning pathways are, the more likely they are to land on the correct solution to these kinds of problems. Variability in the strength of Choose learning pathways, on the other hand, did not relate to solving these problems in any way, shape, or form. To be more specific—this is not to say that carrot learners are bad at solving these types of problems, because carrot learners are not necessarily bad at avoiding bad options. You can be good or bad at both! Instead, the correct way to think about it is that everyone's Avoid accuracy is *more* related to their ability to solve complex problems than their Choose accuracy is. So, why is that?

To better understand the relation between carrot-and-stick learning and problem solving, we built a computer program and taught it to solve the puzzles the same way we thought people might. First, it would pick a visual feature—say, the two curved vertical lines in the top-left image—and then it would try to find a rule that explains how that feature changes in the next block. Afterward it would check to see if it was correct, by testing out the theory on the third block. One *key* to our model was the fact that it needed a way of evaluating itself and deciding whether it was making progress—or moving closer to a solution to the problem. This is often the case in real life. Unfortunately—or maybe fortunately?—we don't often get a big red "INCORRECT" message when we make a suboptimal decision in real life. So part of the trick to solving these complicated problems is figuring out whether what you're doing is working. And unlike many AIs, ours had a way of providing feedback that helped it learn both by carrot (this is working, hooray!) and by stick (you need to forget about this feature). Like the participants in our study, when we improved the model's ability to learn "by stick," it got better, and when we turned up its ability to learn "by carrot," nothing really happened.

Like the data we collected from our participants, our model suggested that when you're in a *complex* problem space, it's important to know when your train of thought is on the wrong track. For instance, if you tried to solve the puzzle by relating the two curved black lines in the top-left image to the three curved black lines in the middle image on the top—say, by a "rotate and add one" rule—you'd be on the wrong track. The solution for this puzzle involves separate rules for the vertical and horizontal lines—any similarities observed between them is a red herring in this problem-solving exercise!

If I'm being completely honest, the fact that only stick learning related to problem-solving success kind of bothered me at first. Though I really believe, from a scientific point of view, that many of our different ways of thinking, feeling, and behaving are associated with both strengths and weaknesses, the idea that "carrot learning" (which I identify with) doesn't have anything to do with problem-solving ability, while moving through life avoiding stairs does, didn't sit quite right with me. However, there are a few noteworthy points that the rider part of my navigation—the part that identifies as an extraverted optimist who is much more driven by the quest for joy than by avoiding disappointment—temporarily forgot.

The first is that most of the choices we make to move toward the good or away from the bad happen subconsciously, or at least at a level that is hard to verbalize. This means that being an optimistic person, or being someone who identifies as being sensitive to rewards or punishments, doesn't necessarily relate to whether you are a carrot or stick learner.

And then there's the fact that carrot learners tend to learn *faster* and more accurately about where to find the best things in life. Carrot learning is powerful. Remember that most of the AIs that use reinforcement learning rely *only* on carrot learning.

But true to the spirit of this book, there are also measurable costs of learning only by carrot. One of the most salient of these costs has been highlighted during this pandemic. When you only have bad options to choose from, carrot learners do *not* shine. In fact, a consequence of their tendency to gravitate toward good options is that they don't learn much about the less-good ones. In other words, you need a stick learner on your zombie apocalypse or pandemic team. You just do.[12]

But here's the deal—under many circumstances, the carrot- and stick-learning systems both converge on the same action. This means that both carrot and stick learners will often end up in the same place—choosing the action with the best chance of a good outcome, even if the experiences that drive them there are completely different.

The moral of the first half of this chapter is that the vast majority of the automatic, intuitive decision-making processes that drive you

---

12. My friend Kristy is my go-to stick learner. She literally saved my life once by screaming so loud she stopped a car from plowing me down in the road. Meanwhile, I was so thoroughly focused on crossing the street to get ice cream, I didn't even notice I was about to die. True story.

through life are reward-based. And whether *your* horse is driven to avoid the least-good things in life or to seek the best ones, without a rider, they are singularly focused on finding their way to rewards.

Thankfully, your horse has a rider. So the critical question then becomes: Can I save you from the pain of public embarrassment by explaining to you what happens when *most people* sing Beyoncé in public? If explicit information can't at least supplement your life experiences, why on earth would I spend *thousands* of hours writing this book or training graduate students? This question brings us back, full circle, to the relation between knowing and doing, and *why* knowing is only half the battle. To finish our discussion, we'll need to complete the description of the battle itself by talking about how the rider of your brain uses what *they* know to influence the way you think, feel, and behave in the world.

Rider navigation: Using conscious recollection to guide your decisions

At last it's time to turn our attention to how your "rider" navigates. After all, they're the ones who decide not only where you want to go but what color hat you're going to be wearing when you get there. And this is the type of navigation that you most likely identify with, since your conscious awareness is what the rider uses to navigate. Because it's in this control space in your flashy prefrontal cortex that you form goals, based on your explicit ideas about what it means to "do better." But what are these ideas made of, and how can they be used to convince your horse of the perils of ice cream?

Throughout this book, I've hinted about what a powerful tool language is. It allows humans to leapfrog over many of our slower, evolutionarily ancient learning systems by following the instructions of others. "Do better," Maya Angelou says, and it makes you *want* to try, even if you're learning about the obstacles that make riding that horse difficult for your brain.

But to actually understand *how* language can be used to guide behaviors, let's have a look inside the metaphorical rider's saddlebag, to see what tools they use to navigate. In it, you'll find the kinds of knowledge that you *can* use language to describe—your declarative memories. For example, I know that an octopus has eight tentacles and can squeeze through any hole larger than its beak; that under certain extreme circumstances, an adult jellyfish can turn itself back into a polyp; that George Washington was the first president of the United States of America; and that two plus two equals four. These "fun facts" make up a subtype of declarative memory called *semantic* memory. Like Wikipedia entries with extensive links to one another, your semantic memories form the web of knowledge that your rider uses to consciously reason about what to do next based on the things they encounter along their path. Among the links you can "click on" in your brain's knowledge web are the meanings of every word you know and can use for describing things to others.

But there are other, *richer* ways of knowing that shape not only how we might answer trivia questions but also how we come to understand our place in the world. In my brain, for instance, knowing that George Washington was the first POTUS looks fundamentally different from knowing that jellyfish can potentially be immortal. This is because I remember the exact moment I first learned about jellyfish. Though most of the details of that specific trip to the aquarium have faded over the past decade, I can still vividly recall what I was looking at (the jellies swimming around a cylindrical tank), what I was thinking about (immortality), and how I felt when my step-mom asked a funny follow-up question about jellyfish (confused, then amused). This specific type of embodied, contextually rich declarative memory is called *episodic memory*.

With respect to your conscious awareness, episodic memories feel like a form of mental time travel that transports you back to the time and place of the original experience, perceived through your rider's first-person perspective. It's like having a series of video clips in which your favorite drama—*The Episodes of You*—is recorded.[13] To access them, you click through files with catchy titles like "First Kiss," "The Jellyfish Question," and "The Stair-Triking Incident."

You might imagine that at each fork in the road of your life, the rider can scan through the declarative memories in their saddlebag for information that might help them decide what to do next. A search through their semantic memory retrieves a pattern match for some of the objects in the world around them: What are these things for? If I eat them, will they increase my blood sugar? Do they provide clues for what lies ahead? Meanwhile, from the episodic database, the rider attempts to retrieve memories of similar events: Have I ever been here before or done anything like this? If so, what can I apply from that experience to this new situation? But in order for what you have learned to be helpful in guiding your decision about what to do next, you need to be able to *find* the relevant information in your ginormous neural database. We'll discuss how this is accomplished, and what happens when it goes wrong, in the next section.

What goes in and what comes out: Memory encoding and retrieval in your brain

So here you are at the next phase of your journey, reaching into your saddlebag of knowledge to fish out a piece of intel that will help you decide what to do next. But you know that *thing* that happens when

---

13. Of course, your real episodic memories have information about touch, smell, and your emotions, which can't be captured in a video clip. Can you imagine what it would be like if we had the technology to record our memories externally with that kind of information?

you're trying to remember the name of a person, a restaurant you liked, or a song title, and you can feel the word forming in your mouth—but you can't get it out? It's like your fingers are grazing the tip of the object in your saddlebag but you can't quite grasp it. "I know I'm in the right filing cabinet," my grandma used to say in the middle of a story about someone whose name starts with *M*, displaying an intuitive understanding that something about how the word was *spelled* was important to where it was located in her brain. The good news is that these "tip-of-the-tongue" moments happen to some extent to everyone. They are a normal[14] by-product of your memory-retrieval processes. The bad news is that they get worse with age, and when you are under stress. And those of us who are "afflicted" with one or both of these conditions can tell you firsthand that *remembering* is definitely part of the other half of the battle.

So what do these tip-of-the-tongue episodes tell us about how our memories are organized? In "Adapt," I used the metaphor of a walk down the beach to help explain the way our experiences change our brains by moving millions of "pieces" into new positions. One of the biological truths this metaphor hints at is that, in the beginning, all memories look more or less the same.[15] Both semantic and episodic memories are essentially woven from the fabric of changing connectivity between neurons.

But to bring either of these types of memory to mind, your brain needs to re-create—with varying degrees of approximation—the

---

14. Of course, as with most things, there is a range in which most phenomena can be considered normal (or abnormal). It's important to note that these tip-of-the-tongue events mostly occur with proper names—not with everyday items like coffee cups or remote controls. Failure to retrieve those kinds of labels may reflect one of the many forms of dementia that also occur in aging.

15. What I mean by "more or less" is that each of our mental experiences corresponds to the coordinated firing patterns of neurons distributed throughout the brain. The exact network changes with the nature of the thought, and with the intensity of your attentional focus, but suffice to say that they aren't apples and oranges.

constellation of neural activity associated with your original experience or experiences. Initially, we can retrace our footsteps with pretty good fidelity. This means that once upon a time, I *probably* had an episodic memory of learning about George Washington. For example, when I was very little, my uncle Percy used to give me a dollar every time I'd visit him.[16] I'm almost positive he told me that the person on the dollar bill was George Washington, and that he was the first president, but at the time I had no idea what a president even was. But that memory would get a little boost each time I saw a dollar bill. Then, in elementary school, I'm sure I learned the very vanilla version of who George Washington was and what role he played in American history. But as the winds of time eroded these events and new memory tracks occluded the original path, the details faded. What remained was a piece of knowledge carved out of the places where several different George Washington–related "walks" overlapped.

To understand how this works, let's revisit some of the principles of learning and neural networks, in the context of how you navigate. The first thing to note is that most of our conscious mental experiences correspond with the synchronized firing of neurons all over the brain, each of which had the job of processing some specific aspect of that experience. During your hypothetical high-blood-sugar diagnosis, for example, a cluster of neurons in the back of your right hemisphere[17] might be focused on your doctor's face, sending that information up into temporo-parietal regions so that you could try to reverse-engineer your doctor's thoughts based on their expression.[18] Meanwhile, your left temporal lobe might be processing the words your doctor was saying, and trying to figure out what they mean. And then there are a whole host of regions we haven't talked

---

16. I told him I would save it for a horse, but it took me thirty years to get there.
17. Assuming the most typical pattern of laterality . . .
18. We'll talk more about these critical social processes in "Connect."

much about, like the amygdala, which kicks into gear when you're afraid.

As you may remember from "Adapt," Hebbian learning tells us that neurons that fire together wire together. As a result, your doctor's-office experience would strengthen the connectivity between the various neural players in this experience. This increases the chance that if some portion of them becomes active again—for example, if you were to see your doctor's face again, or hear the word "diabetes"—the rest of the neurons may also become active and "trigger" a spontaneous retrieval of that memory. This effect in the brain reminds me of those complex domino-stacking patterns that people with more patience than me create. Someone knocks down the first one and it knocks down others, and before you know it, they are falling in elaborate patterns across the table.[19] When two neurons fire together, it's as if you've moved two dominoes on the table closer together. And if you scale this up to the 80-billion-domino level, you have a pretty good model for memory retrieval in the brain.

Of course, the dynamics involved in memory formation and retrieval in your brain are more complicated. Remember the "noisy" communication conditions that drove our brain-design discussions in "Mixology" and "In Sync"? One thing we didn't discuss is that because of the noise in the brain, sometimes a neuron will fire randomly, without any external triggering event. The effect is kind of like having a mouse running around on your table of dominoes, randomly nudging dominoes and even tipping a few over in different places. And of course, each domino in your brain's version might connect to thousands of others.

Then there's the fact that the strength of connections between neurons, or the distance between dominoes, changes dynamically

---

19. If you want to see something fun that approximates your brain's scale of complexity, search for "1,000,000 Dominoes Falling is Oddly SATISFYING" (by Hevesh5) on YouTube.

with each new experience. This memory modification occurs through two processes: decay and interference. Sometimes, when two neurons that fired and wired together simply don't fire together again, the connections between them weaken, and decay-based forgetting occurs. The effect of this is like removing one of the dominoes in the chain. Depending on how many dominoes get removed, where in the chain they were, and how far away its neighbors were, such forgetting can either blur the details of a memory or completely wipe it away. This can also create a "tip-of-the-tongue" phenomenon. It's like your brain has expertly knocked down the "M" domino, but the domino tipping stopped before the rest of the pattern could be retrieved.

Interference, on the other hand, occurs when one of the two neurons that fired together also fires and wires with someone else. The effect is like placing two dominoes side by side in the chain, each with different downstream consequences. Depending on how many times this happens, and how elaborate the downstream effects are, you may no longer be able to recognize the original pattern in the new memory. A perfect real-world example of this is the frustrating experience of trying to remember where you parked your car, especially in places where you park often, like the grocery store. After dozens of similar events with slightly different details encoded, it can be *really hard* to retrieve the pattern that corresponds to one specific parking event!

Together, the processes of decay and interference work in tandem to shape the memories you form. And though both episodic and semantic memories are modified by these processes, your episodic memories are particularly vulnerable to them. Since many of the moments of your daily life, like parking at the grocery store, have lots of overlap in their details, episodic memories often fall victim to interference. To remember a particular event, like where you parked your car or what you wore on Thursday, you need to be able to bind the people, objects, and actions involved to a specific place and time in the reactivation process.

And *this* is where one of the biggest differences between semantic and episodic memory comes into play. The ability to re-create a pattern of neural activity that includes the contextual details that distinguish one of the episodes of your life from the next requires a very specific brain region—the hippocampus. You were first introduced to this seahorse-shaped brain area in "Introductions." As you might remember,[20] it's the region that changes shape when London cab drivers memorize tens of thousands of different maps. And it's no accident that the part of the brain that supports spatial navigation in humans and other vertebrate animals is also the one that is critically involved in encoding or retrieving any memory that has your *first-person perspective* preserved. In the relatively long period of evolution before Google Maps was invented, learning to navigate through space was accomplished by moving *yourself* around in the world and learning how to get from point A to point B by remembering how the details of the space around you changed.

Critically for how you navigate, there are neurons in the hippocampus, appropriately named *place cells,* that *keep track* of where you are in a particular environment. But contemporary theories of the role of the hippocampus in memory suggest that these place cells might be involved, more generally, in creating *meaning maps* based on our experiences. Here's how this works, in a nutshell: As people (as well as some animals) gain experience moving around in a particular space, they are able to "stitch together" their individual experiences to form a mental map. When they do, they can mentally subtract themselves from the picture, moving from an *egocentric* perspective—one based solely on where things are in respect to

20. It's totally normal if you don't remember this. It was a small detail in a story you read a long time ago. You've encountered a lot of new information in this book along the way!

themselves—to more of a bird's-eye view that represents where things lie in respect to one another, or an *allocentric* perspective. If you can make it from your bedroom to the bathroom, and back, in the dark, without bumping into any furniture, you probably have a pretty accurate mental map of those spaces, and their relative position to one another.[21]

EGOCENTRIC PERSPECTIVE

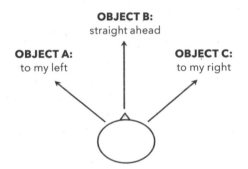

OBJECT B:
straight ahead

OBJECT A:
to my left

OBJECT C:
to my right

ALLOCENTRIC PERSPECTIVE

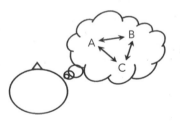

And here's where things get even more interesting, in my opinion. In much the same way that taxi drivers form mental maps about where spatial landmarks exist relative to one another based on their

_____

21. If you take this trip as frequently as I do, however, it's also possible that your horse can navigate you there through procedural memory. The fact that I can do it while half-asleep suggests that this is the case for me.

experiences, your brain also forms maps of the relations between the people, places, and events you encounter along the way.

Let's play a game to show how this works. I'm going to give you a list of words to read, and I want you to say the first word that comes to your mind when you read it.

dog, _____, salt, _____, doctor, _____, coffee, _____

According to a database collected on the responses of over 6,000 English-speakers who did similar "free-association" tests with these words, your responses are likely to be cat (67%), pepper (70%), nurse (38%), and tea (44%).

What does it mean that one word can trigger the same response in so many people? Or perhaps you might be more interested to understand what it says about your brain if you said something different from the norm? In short, unless you did something *funny*—like trying to be creative instead of following directions and saying the first thing that came to mind—the first word that popped into your head is probably the one that has the most connections to the target word in your meaning space.[22] If you think for a second about *why* dogs and cats, salt and pepper, doctors and nurses, and coffee and tea might be neighbors, one thing pops out: You encounter them in similar situations or contexts. Sometimes they coexist, like salt and pepper at mealtime, or doctors and nurses in healthcare settings. And other times, you might find one or another, like cats versus dogs as household pets, or coffee versus tea as a morning routine; but these things *function* roughly the same. If they are stored together in your meaning space, they fill roughly the same niche in the *Episodes of You*.

However, according to a memory model proposed by Charan

---

22. In the next section, we'll talk a bit more about individual differences in the structure of these spaces.

Ranganath and Maureen Ritchey, there are fundamentally different *types* of maps that the hippocampus forms, and these different types of maps are organized according to the connections between the hippocampus and the rest of the brain. In their review of the memory literature, Ranganath and Ritchey describe two systems for guiding memory-based behavior that are driven by different parts of the hippocampus—one that maps familiar people and things, their features, and their relations to one another, and another that understands scenarios or contexts, which can include spatial locations (for example, things that happen at home); temporal locations (for example, things that happen in the morning); and more complex event structures based on their combination (for example, things that happen at home in the morning versus things that happen at home in the evening). Consistent with this idea, a recent experiment by Mladen Sormaz and colleagues found that people's ability to remember information about either semantic knowledge or spatial information is related to the patterns of connectivity between their hippocampi and the rest of their brains.

Sormaz and colleagues' first step was to estimate patterns of connectivity between the hippocampus and the rest of the brain based on task-free functional MRI data collected from 136 participants who were allowed to mind-wander in the scanner. But instead of measuring how much of their brain was speaking over a certain frequency, as we do with EEG data, this study measured the ebb and flow of activation across different regions of each participant's brain while they mind-wandered in the narrow tube. One underlying assumption of these measures is that the degree to which activation in two brain regions rises and falls together reflects how in sync they are, which relates to how frequently coupled their processes are across different situations. These patterns of connectivity were then related to performance on behavioral tests of memory performed outside of the scanner.

One of the most remarkable findings the team reported[23] was a pattern of connectivity in the brain that reliably distinguished people who were good at remembering "what" things are from those who remembered "where" things are. The pattern related to the degree of lateralization, lopsidedness, or connectivity between the left and right hippocampi and a region at the intersection of the left temporal and parietal lobes that is frequently associated with the retrieval of word meanings. Specifically, they found that people who had stronger connectivity between this left temporo-parietal region and the left hippocampus, as compared to the right hippocampus, performed better on a semantic memory task. However, this same pattern of connectivity was associated with poorer performance on a task that measured spatial, or topographic, memory. And the reverse pattern was also true. People who had stronger connectivity between the right hippocampus and the left temporo-parietal junction than between the left hippocampus and this same region performed better on the spatial memory task but worse on the semantic memory tasks. These findings seem to suggest that there is some degree of competition in the brain for being able to remember where things are versus being able to recognize what they are,[24] a phenomenon that is likely tied to the role of the hippocampus in *reinstating* the patterns of

---

23. It's worth mentioning that the authors reported several relations between hippocampal connectivity and memory, but I had to narrow it down in the interest of space!

24. Andrea and I fall squarely into these categories. I'm incredibly good at recognition memory. I have a pretty good vocabulary and am excellent at recognizing faces more specifically (though I'm absolute crap with names). Sometimes we'll be watching a TV show and I'll recognize an actor that played a minor role in a show we watched a decade earlier. Andrea, on the other hand, has an incredible memory for times and places. "Have you seen my glasses?" I might ask him, and he's somehow able to scan through his visual memory and retrieve the event of seeing them in whatever strange place they like to hide themselves. "Do you remember when we opened our checking account, by any chance?" I might ask him (since I know about his incredible abilities). "Yeah, I think it was April or May 2007." Hey—I never said that my skills were as useful as Andrea's . . .

activation associated with each kind of memory when your rider does a search.

While I hope you find this as interesting as I do, it still doesn't explain the problem we started this section with. Why is it so tricky to remember the *names* of people or locations? The answer connects the dots between many of the details you've learned about putting things in memory and getting them back out. To put it simply, proper names, which are at the root of most tip-of-the-tongue experiences, live in a kind of limbo in your memory space—somewhere between episodic and semantic memories. On average, you encounter any specific proper name, like Jasmine, Ringo, Seattle, or Twilight Exit much less frequently than you encounter common names, like daughter, dog, city, or bar. So unless these names identify people, places, or things that are very familiar to you, the paths between the names and their intended referents are *not* well worn. This makes them harder to find. And then there's the problem of interference. How many names and faces do you have mapped together in your database? How many Karens do you know? Add this to the fact that there is a good degree of arbitrariness in proper names. Unlike apples and oranges, which tend to share certain predictable features and occur in the same contexts, the dimensions in your brain that separate the faces of "Karens" from those of "Saras" are much less predictable.[25] When you add all of these factors up, it makes sense that there are times when the patterns corresponding to proper names only get partially retrieved. In the next section, we'll turn our focus to the things we know that are less frustrating to retrieve by digging into the meaning maps your brain creates based on the way things systematically relate to *you*.

---

25. Unless one is calling a manager...

Neuroscience of knowing: Meaning maps in the brain

Over the past fifteen years, the "mind reading" experiments conducted by Marcel Just and colleagues have greatly improved our understanding of the *structure* of the meaning maps our brains use to guide us through our lives. In June 2005, when I first came to Carnegie Mellon to be trained by Marcel, this fascinating body of research was in the midst of taking off. In collaboration with computer scientist Tom Mitchell, my friends Svetlana Shinkareva and Rob Mason, and a team of other brilliant minds, Marcel set out to understand the *physical structure* of our thoughts. Although I was not directly involved in these projects in any way, just being in the room was enough exposure for me to appreciate how complicated the adventure was.

The problem, in a nutshell, is that typical analyses of fMRI data are designed to tell you whether one specific region of the brain is involved in some mental function of interest. To understand the neural underpinnings of Function X, for example, patterns of activation in Brain Region A are explored, to see whether it systematically becomes more active when Function X is happening (ideally, when compared to some well-controlled Function Y). You then repeat the process for Brain Regions B, C, D, and so on, independently. When you're done, you get a map of regions that are more active in Function X than in Function Y. Depending on the differences between the two functions, you can make inferences about what those parts of the brain might be doing. For instance, the areas that are more involved in reading real words, like "drill," than in reading nonsense words, like "blicket," might be involved in the retrieval of semantic information through language.[26]

---

26. Of course, whatever you ask participants to *do* with the words or nonwords also shapes what the brain will consider—but that can be set aside for now.

The challenge is that this type of analysis is not sensitive enough to detect the subtle differences associated with the structure of our semantic knowledge, like the difference between thinking about a hammer versus a drill. One reason lies in the *size* of the individual brain areas that we investigate with these methods. For practical purposes, functional neuroimaging studies tend to carve the brain into areas that are one millimeter cubed, at the smallest. This is roughly the size of the tip of a sharpened pencil. That might seem small, but the signal recorded in each of these areas is driven by the firing of over half a million neurons. The result is like listening to a neighborhood with hundreds of thousands of gossipy neighbors. In traditional analyses, you're only interested in whether the collective noise of the neighborhood gets louder or not. But with these "neurosemantic" studies, the goal is to understand when their conversation changes topic. Within a given neighborhood, individual neurons might respond differently to the idea of a hammer versus a drill, but if the *number* of neurons responding isn't very different, traditional analyses won't be able to detect the change.

The key to solving this problem is to recognize that semantic knowledge doesn't live in one neighborhood alone. To determine whether a different group of neurons in Neighborhood A is responding to the words "drill" and "hammer," we need to figure out who else they're talking to. This means we need a way to investigate how activation in Neighborhoods B, C, D, and so on also changes when the participant is viewing a drill or a hammer. In 2001, James Haxby and his team developed a technique called multivoxel pattern analysis (MVPA) that does precisely this, and with it, Just and colleagues set out to uncover the way the human brain maps meaning.

In the first of many neurosemantics studies, the team—led by Svetlana Shinkareva—showed participants ten different line drawings. Five were images of tools (hammer, drill, screwdriver, pliers, and saw) and five were of dwellings (house, apartment, castle, hut, and igloo). To train people to consistently activate the web of knowledge associated with each drawing, the experimenters showed their participants pictures outside of the scanner and had them practice thinking about their properties. What would it feel like if you touched it or held it in your hand? What would you use it for? Where would you see it? Then, in the scanner, study participants saw each image six times, in varying orders, while their patterns of brain activation were recorded.

The goal of the research team was to try to figure out what a person was thinking about, based on their distributed patterns of brain activation. If they couldn't detect the exact image, could they at least figure out whether the person was thinking about a tool or a dwelling? To "read the minds" of their participants based on their brain activation, Shinkareva and colleagues combined neuroscience methods with the tools of computer science. Using a technique called *machine learning*—a name given to computer algorithms that learn, like we do, through examples—the experimenters fed patterns of brain data into a computer algorithm called a "classifier." The classifier's job is to learn how to identify what group, or "class," a string of data it is fed belongs to. For example, the computer might receive one hundred values corresponding to the amount of activation in one hundred brain regions when a person saw the image of a hammer, along with the label "hammer." Then it got another one hundred values corresponding to activation in the same one hundred regions when a drill was viewed, followed by the label "drill." As the training set grew in size, the classifier learned to identify the pattern of activation across the one hundred values[27] that was associated with each image. And it learned remarkably well. After receiving fifty data sets, with the patterns of brain activation corresponding to the ten drawings

---

27. This number is just an example. The actual number of brain regions varied across the many different analyses that were conducted.

viewed five times each, it was fed a string of data it hadn't seen before and was asked to guess what it was. In the best participant, it was able to guess the exact item they were thinking about with 94 percent accuracy. In the worst, it was closer to 60 percent accuracy. We'll return to what these differences between participants might reflect in a minute.

And if the ability to detect what someone was thinking based on their *own* brain data wasn't remarkable enough, Shinkareva and team took it a step further and showed that they could predict what one person was thinking when they trained the classifier on the brain activity of the *other* people in the study. As you might suspect based on everything you've learned in this book so far, these classifiers performed worse, on average.[28] This makes sense if you consider the fact that people have different experiences with drills and hammers, and that classifiers trained on their own data would have more information about their unique perspectives. In spite of these differences, it's pretty amazing that this between-person approach worked for most participants. In fact, Shinkareva and colleagues were able to predict what 75 percent of their participants were thinking based on the data of others. Of course, this raises the question: What was different about that other 25 percent? Taken together, the results from this study highlight that there are both commonalities and differences in the way semantic information gets mapped in the brain across individuals.

Since their groundbreaking study, Just and collaborators have conducted dozens of investigations of the way the human mind maps meaning. In one study, for instance, they investigated patterns of activation in sixty concrete objects, adding categories like food, animals, and vehicles to the tools-and-dwellings mix. They then grouped

---

28. The best of these classifiers performed with an accuracy of about 80 percent. It is worth noting, however, that for two participants, the classifier was better at predicting what they were looking at using other people's data than their own!

the objects based on how similar their patterns of activation were and deduced that, when it comes to how the brain represents knowledge about objects, there are three major organizational themes: (1) Can I eat it, or is it involved in eating?[29] (2) Can I hold it in my hand or manipulate it with my hand? And (3) Can I go inside it, or use it for shelter?

These organization principles make good sense when you remember that semantic knowledge can stem from the commonalities in our episodic memories. They also explain something about why *your* patterns of brain activity might be useful in determining what *I* am thinking about. The chances are excellent that although you and I might have different experiences with celery, we both associate it with cooking or eating. I'd even take it a step further and say that neither of us holds our celery like a pencil or eats it like corn on the cob.[30] So what might this tell us about the people whose thoughts are harder to classify, using either their own data or the data of others?

One important thing to note is that the ability of a classifier to discriminate one thought from another relates to how *distinct* the neural representations are in the mind of an individual.[31] And a person's experience with the items being classified shapes this distinction. For example, the concepts of coffee and tea *must* be more different in my brain than they are in the brain of someone who drinks neither, based on our experiences. While *they* might see interchangeable, frequently caffeinated, typically brown beverages that are consumed in cups with handles, *I* see Italy versus the U.K., morn-

---

29. Utensils, like cups, and animals commonly eaten in the predominant culture of these participants, like cows, looked more similar to food items, like carrots, than they did to other tools or animals that aren't involved in eating scenarios.

30. Though, if you are this kind of weird, we could definitely be friends.

31. As well as to how noisy or variable their measurements are—but we'll set that aside for now.

ing versus afternoon, hot versus cold, and health versus illness. You might predict, then, that a classifier would be more likely to tell whether *I'm* thinking about coffee or tea than it would be based on the brain data of the hypothetical "they who don't drink caffeinated beverages." You might also (correctly) predict that "tea" is not the first word that comes to my mind when presented with the word "coffee."[32] Can you reverse-engineer how your experiences might have driven the associations you came up with?[33]

In "Adapt," we described some of the serious ways your experiences might shape the similarities in your semantic neighborhoods—by forming contextual links between Black faces and weapons, for instance. A study published in 2017 by Marcel Just's team demonstrated another chilling implication of the way our experiences shape our meaning maps. Their participants included a group of individuals who reported having suicidal thoughts, and another group of individuals who did not report such thoughts.[34] This time, the goal of the research team was to see if the classifier could learn to identify something about the *thinker* of the thoughts, based on the patterns of activation their brains produced in response to words. But instead of looking at tools, dwellings, or other objects, the participants in this study were given more abstract concepts to think about, which related to either negative thoughts (for example, death, hopeless, desperate) or positive ones (for example, bliss, carefree, kindness). Astonishingly, based on the patterns of brain activation generated by these concepts, the classifier was able to detect with 91 percent accuracy whether the thinker of the thought was from the suicidal

---

32. Instead, I'm one of the 8 percent who think "caffeine" when I see the word.

33. Warning: It's important to note that when you do so, you're using your rider to infer a purely horse-driven process.

34. These individuals were selected based on assessments that showed they were not having current mental-health challenges and did not have any history of suicidal ideation.

group or the mentally healthy group. The difference between groups was apparent in their patterns of activation for *both* strongly negative and positive words, with the largest group differences observed when people thought about death, cruelty, trouble, carefree, good, and praise.

There is a lot to take away from this experiment. From a detached, scientific standpoint, one thing to note is that the way we *feel* during our experiences also shapes the way our memories become organized. But if you view these results through a less clinical, more humane lens, you might wonder—if the patterns of brain activity of a suicidal person thinking about the word "good" are distinct enough that a *computer* can tell them apart from someone in a healthier mindset, how different must their *experiences* of the world be? And finally, when the fundamental building blocks on which more complex ideas are based look so different in the minds of people with different experiences, how can we find a way to realign and connect?[35] These are critical questions that I hope neuroscientists and clinicians will work together to address, to better help the more than 264 million people worldwide who are affected by depression.[36]

But one of the known challenges to treating depression is that our emotional states can also influence what we notice, or pay attention to, in any given experience. And whatever we focus on gets magnified in the way our memories are stored. This might not only drive how distinct two concepts are in a person's brain but also *where* in their brains the distinctions are stored. One hot-off-the-presses study, published in 2021 by Katherine Alfred and colleagues, showed precisely this—that systematic differences in what people pay attention to drive the way their brains come to represent meaning.

---

35. This will be the focus of the final chapter!
36. And these statistics, provided by the World Health Organization, are based on pre-pandemic data!

To study this, Alfred and colleagues developed a clever test designed to measure whether a person is more likely to focus on verbal or visual information. In it, they presented a series of black-and-white stimuli that resembled playing cards. Each one had one of three shapes commonly found on a deck of cards—a heart, a club, or a spade. But instead of numbers, these cards also had the *word*—"heart," "club," or "spade"—printed either above or below the shape.[37] The participants were asked to press one of three buttons to "sort" the cards into their three respective suits. Most of the time, the shape and the word matched, so people could use either piece of information to sort the cards. But every once in a while, a "trick" card would be presented, with inconsistent word-shape pairings. For example, the word "heart" might be printed above the shape of a spade. Participants were neither warned about these trick cards nor told how to sort them. The experimenters hoped to figure out what kind of information people were more focused on—visual or verbal—by recording the decisions they made when conflicting information was presented in the two modalities.

People's performance on the card-sorting task showed that almost everyone had a predisposition to attend to either verbal or visual information. Some people sorted trick cards almost entirely based on the words presented, while others sorted them primarily based on pictures. By subtracting the number of times each participant sorted fifty trick cards based on their pictures from the number of times they sorted them based on words, the experimenters calculated a "word-bias" score that ranged from +50 to –50.

Next, the researchers wanted to see whether these attentional

---

37. I haven't a clue why diamonds were left out of this task, but it might be because the word "diamond" is much longer than the others and might have "grabbed" attention accordingly. The scientists were very careful to control for things that can influence attention, like the position of the information on the card, by shifting the relative positions of the word or picture between trials.

biases were related to the way a person's brain represented the meanings of concrete objects. To do this, they recorded patterns of brain activity while their participants viewed sixty objects in both word and picture formats. Then, they used a variation of MVPA called the *searchlight method* to explore how semantic information was represented in each person's brain. You might think of this as a hybrid between the traditional neuroimaging analyses that focus on activation in one brain area at a time and the MVPA that looks for patterns distributed across the whole brain at once. As the name suggests, "searchlights" look for patterns of activation in brain regions that are next to each other in a predefined physical search space. The goal of this method is usually to see how classification accuracy changes when the searchlight changes location.

But Alfred and colleagues took it one step further. They wanted to see if there were parts of the brain in which *individuals* showed differences in the way their thoughts about objects were organized, based on their attentional biases. Using the searchlight method, the researchers moved from one brain region to the next, relating a person's tendency to pay attention to pictures or words to the *meaning map* their brains created based on the relations between those sixty objects.[38] Their results uncovered something remarkable: There were three brain regions in which word-focused participants and picture-focused participants showed fundamental differences in the way their meaning maps were organized. One of these regions is the left temporo-parietal junction, the same region whose connectivity to the hippocampus reliably distinguished people who were good at remembering "what" from those who were good at remembering

---

38. I simplified the analysis a bit here to save time and mental energy for those who might not be interested in the nitty-gritty details. If you are interested, I'd encourage you to check out their publication, which explains how they used a sophisticated algorithm to compute the semantic distance between the objects in their experiment (presented both in words and pictures) and used that as a template to relate patterns of activation to.

"where." The same region has been shown to be more active in people who self-identify as having verbal processing styles.

In other words,[39] these studies suggest that people who tend to focus more on verbal or semantic information have sharper, or more distinct, meaning maps in one of the areas typically associated with language in the left hemisphere. They also have stronger connections between the left hippocampus and left-hemisphere language regions, which would make it easier for them to *retrieve* or reactivate concepts based on the meaning maps stored in these regions. But for people who tend to focus on scenes and images, the representations in these areas are less distinct, and their connections to the hippocampus in the left hemisphere are less strong, which may make it harder for them to retrieve semantic concepts based on words. Perhaps this explains why some people *don't* think by "talking to themselves" and others find it difficult to conjure images in their mind—we all lean into the codes that our brains find the most efficient for representing the world "out there."

At the end of the day, the whole host of processes that guides memory encoding, storage, and retrieval shapes the way the rider in your brain navigates. Because the extent to which anything you "know" can be used to "do better" depends on your ability to *retrieve* that memory at the appropriate time and place, and to use it to guide you.

But what about all of the "fun facts" we know that seem irrelevant to any of our real decision-making spaces? In the pursuit of my day job, exploring basic[40] scientific questions about how brains work, I often find myself identifying with Tyrion Lannister's view on

---

39. Pun intended.

40. I know it sounds a bit condescending, but "basic" science is a term used to describe research that tries to understand the fundamental mechanisms of how things work, as opposed to solving clinical or applied problems.

knowing, which is somewhat different in spirit from Maya Angelou's.[41] "That's what I do," he says. "I drink and I know things." And when he says this, it makes me feel a little bit better about the satisfaction that knowing, just for the sake of knowing, can bring. In the next chapter, you'll learn *why* some of us find it extremely satisfying to learn something new, even if it is completely unlikely to help us *do* anything better. But before we get there, let's wrap up what we've learned about how different horse and rider control systems go about using what they've learned in life to navigate.

Summary: The horse and rider "live and learn" in unique ways that jointly guide you through life

I hope that after reading this chapter, you have a richer understanding of the relationship between knowing and doing. Is knowing *really* only half the battle? If so, what gets in the way of doing better once you know something that could improve your behaviors? One thing to keep in mind is that, as we discussed, your brain has different *ways* of knowing. And as I mentioned briefly in "Adapt," it is perfectly possible for your horse's automatic and intuitive way of knowing how to navigate to be at odds with your rider's conscious, explicit goals and ideals for how you should behave. And when this happens, the battle between horse and rider plays out in much the same way that the competition for focus does. The extent to which any decision is driven by the horse, the rider, or a combination of the two depends a lot on the relative strength of the two processes in your brain. And while the rider can often use a piece of information they've retrieved to turn the horse in a different direction, they are also much quicker to tire than the horse is.

---

41. Tyrion Lannister (from *Game of Thrones*) is one of the fictional characters I would most enjoy hanging out with.

But here's something we didn't talk much about before: With *practice*, navigating in a way that starts out effortful and rider-driven can become a more automatic, horse-driven task. Remember learning to drive a car? Most people don't stick their teenager into a car and say, "Point it toward ice cream and try not to die." Instead, we give them a list of explicit instructions: "Adjust the mirrors and your seat. Place your hands at two and ten o'clock. Check the mirrors and your blind spot before changing lanes" and so on. And because this is a lot of information for the rider to hold in mind, we are there (or we find someone braver to be there) to remind them if they forget. But at the end of the day, no one gets good at driving a car without practicing. Eventually, all of these tasks, initially encoded through verbal instructions, become so well practiced that your automatic control systems can easily navigate you through traffic without ever going through "the checklist." Your horse *can* learn a new trick, and when it does, it forms a new set of associations between actions and rewards in your brain. Keep this in mind if you are motivated to change something and you feel like you're struggling with your horse. Practice might not make you perfect, but it certainly makes things less effortful.[42]

In this chapter, we also learned about individual differences that happen *within* the horse and rider control systems. Are you a carrot learner, a stick learner, or both? Does your brain learn more when things go worse than expected, or when they go better than expected? How might this help you figure out whether you're on the right track when you're trying to do something complicated?

And finally, we learned about the tools your rider can use to navigate—or the contents of their saddlebag. Inside are the elaborate clips of memory that form the *Episodes of You*, along with the truisms

---

42. Also keep in mind that the rider is more sensitive to things like stress and fatigue, because that type of control is incredibly energetically demanding in the brain.

and fun facts that you come to acquire through repeated experiences. How do the ways that you focus interact with your life experiences to shape the meaning maps in your brain? And how do the connections formed influence what your rider will find when they reach into their bag to pull out memories that might help guide their next steps?

But now that we've covered this bit of ground together, the space where different types of knowing influence what we do, I'm going to complicate things a bit by reminding you that sometimes we *like* to know things—like the fact that an octopus can squeeze through a tiny hole—even if there is very little possibility for these things to change our behaviors in the future. Where does that come from? In the next chapter, we'll talk about different ways brains respond to the unknown, and why some brains are more motivated than others to know things just for the sake of knowing.

CHAPTER 7

EXPLORE

How Curiosity and Threat Compete to Shape
Behaviors at the Edges of Knowing

"What's the *point* of a jellyfish?"

I'd like to think I've been asked some tough questions in my life, but this one takes the cake. Fortunately, it was directed to Jasmine, our resident marine-life expert, and *not* to me. The question was asked by my stepmom, Linda, one of the most playful and adventurous adults I know, during a visit to the Seattle Aquarium. Jasmine, who had worked at the aquarium for several years, was playing the role of our personal tour guide—and man, was she good at it. Though she didn't begin studying marine biology formally until high school, Jasmine has been fascinated by the aquatic world since she first opened her eyes underwater. And by the time she reached her teens, she was a Pez dispenser full of "fun facts" about the various sea critters we encountered.

At the moment Linda's question dropped, I was deeply engrossed in mind wandering, rolling the piece of information I had just learned about jellyfish around and around in my head. As I mentioned briefly in the last chapter, I had just learned that some species of jellyfish are capable of transforming from their adult form—the umbrella with legs, more formally known as a medusa—into an immobile polyp form, which typically occurs much *earlier* in their life cycle.

What?

This is like the invertebrate equivalent of saying that a chicken can turn *back* into an egg for a while if it gets injured or can't find food! The mere possibility that such a thing could happen defies everything I *thought* I understood about how living things worked. So while my eyes watched the hypnotic movements of the jellies in their medusa forms, my mind tried to grapple with the possibility that one of them could live forever.

When "What's the *point* of a jellyfish?" landed on my eardrums, I was so far down the immortality rabbit hole that I almost couldn't *understand* the question. Linda's words, uttered by a brain in a very different state of wondering, was so unexpected, it created a kind of thought whiplash that rendered me temporarily utterly confused. It was like seeing the guy in my neighborhood walking goats, multiplied by a thousand. Her question was so unexpected, she effectively stumped all three science-minded folks in the group by asking us a very *practical* question about jellyfish.

I still have *no idea* how to answer Linda's question, but I do know that *the point* of the story is to illustrate the different ways people think, feel, and behave when they encounter a new piece of information or an unexpected situation, and how this might relate to how useful we estimate this information to be in the real world. On the one hand, you have Jasmine, whose early fascination with marine animals has been a huge driving force in her life. From her first volunteer position at the Seattle Aquarium in her early teens to her current job working for the National Oceanic and Atmospheric Administration (NOAA) on the policies that govern fishing practices worldwide, Jasmine's life has been significantly and consistently shaped by her curiosity about marine life. On the other hand, you have Linda, whose area of expertise might best be described as "fun!" As a result, Linda has lived a life full of adventure and has many entertaining stories to tell about it. I lie somewhere between them, along with Jimmy Buffett,[1] wondering whether it would be any fun to *be* a jellyfish.

Remarkably, the relation between how you explore the unknown and the map you will build to navigate the world kind of resembles the life cycle of a jellyfish. The fact that I still *remember* that jellyfish can turn back the hands of time on their life cycle, though I have categorically forgotten every other "fun fact" presented to me on that trip, provides real-world evidence for what neuroscientists are now able to demonstrate in the lab. Curiosity is a mental state that both precedes and facilitates learning. Put simply, curiosity is the subjective feeling one gets when their brain wants to take in a piece of information in front of them. As a result, the more curious you feel in any given situation, the more prepared your brain is to *remember* what happens next.

The way that a brain, hungry for information, can drive you to explore the unknown parts of your world is observable from a very early point in life. The research of my friend and former colleague Kelsey Lucca has demonstrated this repeatedly in her studies of spontaneous pointing gestures in infants and toddlers. Kelsey and her collaborators have shown that if you name a new object when an eighteen-month-old toddler[2] points to it, they are more likely to *remember* the object's name later. To demonstrate this in the lab, they compared situations like this to two others—one in which the experimenters named something when the toddlers weren't pointing to

---

1. Setting aside the fact that I am a huge fan of most European accents, some of the lyrics of Jimmy Buffett's song "Mental Floss" really call to me.

2. By comparison, there was no effect of naming objects when twelve-month-olds pointed to them.

anything, ostensibly a sign that they weren't very interested in the new object, and another in which they named something *different* from the object the toddlers were pointing to. In both conditions, babies were less likely to remember the names provided than they were when given the name to something they had expressed interest in. These results suggest that *pointing* is a clever tool that the toddler brain has developed for asking a question about what something is before it has the words to do so. The notion that a person's curiosity can be pointed toward a *specific* target, and that such curiosity fosters learning, has also been demonstrated in adults. One common way to study this is by using modified trivia games in the lab. In these experiments, participants read a series of questions designed to elicit curiosity in people with diverse interests: What is Quentin Tarantino's favorite movie? Which instrument was designed to mimic the sound of a human voice? How many NBA championships did Michael Jordan win with the Chicago Bulls? How many tattoos does Post Malone have?[3] After reading each question, participants are asked to rate both how *confident* they are that they already know the answer to the question and how *curious* they are to learn the answer. Then, most of the time, the answers to the questions are provided.[4] Like the pointing toddlers, adults are more likely to *remember* the answers to the questions they were most curious about when given a pop quiz at the end of the experiment.

Of course, this raises the question *why* one person would be interested in learning about Post Malone's tattoos, while another would be more curious about Michael Jordan's winning streak with the

---

3. I don't know if anyone knows the answers to the last question besides Post Malone, because I totally made it up to pique your curiosity—but the others are taken from actual research paradigms. The answers to the first three questions, based on these experiments (so don't blame me if Quentin Tarantino changed his favorite movie) are: *Battle Royale*, violin, and six.

4. We'll get into the more nuanced details of the design of these experiments, and the reasons for them, later in the chapter.

Bulls. And this is where the cycle of questioning and answering starts to look like the immortal life of a jellyfish. According to the Prediction, Appraisal, Curiosity, and Exploration (PACE) framework recently developed by Matthias Gruber and his former mentor Charan Ranganath,[5] your curiosity in any given situation depends on *what you already know* about the world. Put simply, your curiosity gets piqued when something either surprises you based on what you *thought* you knew[6] or because you experience a knowledge gap—a type of mental conflict that occurs when you need more information before deciding what to do in a given situation.[7]

Take, for example, the "If you're having a bad day just look at this shaved llama" meme that recently hit the circuit. While I am pretty sure that most people who saw it were most captivated by the hysterical, pissed-off expression on the animal's face, or the fact that its head looked like a dandelion, what surprised *me* about the meme was that I was pretty sure that the shaved llama was an alpaca. My curiosity about this inconsistency drove me to the Internet, where I confirmed what I had learned at the county fair a few years ago about the difference between the two. This sent me down yet another rabbit hole of wondering, where I learned about the relative ease of domesticating llamas and alpacas. It turns out that llamas are more friendly

---

5. It's not your imagination. This is the third time you've heard about Charan Ranganath, and it's not the last. I should probably set aside some of the proceeds from my book for him, since his research is so central to it! I was fortunate enough to learn from him when he was first hired as an assistant professor at UC Davis during my graduate training, and he's as brilliant and funny as you might hope someone who studies curiosity might be.

6. This is why experimenters try to account for previous knowledge by asking people how confident they are that they already know the answer to a question. As you'll read later in this chapter, surprise can drive learning as much as, or more than, pure interest.

7. For instance, if you encountered the name of a person you didn't already know in one of the example trivia questions, you might have felt motivated to do a quick Internet search to figure out who it was before deciding whether you were curious about them. (Please don't tell me if it was Michael Jordan.)

and doglike, while alpacas are more independent and catlike. But they're *both* really funny to look at—llamas with their long, derpy ears and noses, and alpacas with their snubby, button-like faces. As my database of knowledge increases, the space in which *either* llama *or* alpaca memes can capture my interest grows larger by the minute.

Caught in this cycle of wondering and knowing, one can iterate their way through life, feeling more curious, and possibly even more clueless, with each new day. I *believe* this was the idea that Plato tried to capture when he described the irony of his teacher, Socrates, and his attitude about knowing. Though Socrates is frequently considered to be one of the wisest people who ever lived, he famously claimed to "neither know nor think that I know." But what about those of you who are more practical—like Linda—and might *not* be captivated by the idea of knowing for the sake of knowing? Are you missing out on the opportunity to feel as "wise" as Socrates was?

Hopefully, you know me well enough by now to guess that it's never that simple. As you'll learn in this chapter, there can be significant costs to exploring the unknown. These costs range from "wasting time" on the low end to the possibility of discovering something that can harm you physically or psychologically on the high end.

So what's the *point* of exploring new places or ideas?

To answer this question, we'll return to the PACE model and the cycle of questioning and answering, with the costs and benefits of this pursuit in mind. But before we do, let's do a little assessment to figure out how strong your desire to know for the sake of knowing is.

How curious are you by nature?

Returning to the Greek philosophers for a minute, I'd like to begin our deeper dive into individual differences in curiosity the same way that Aristotle's collection *Metaphysics* begins—with speculation about the *nature* of human curiosity. "ALL humans by nature desire to know," the opening sentence boldly claims. But I can't help but wonder whether his beliefs on the topic were biased by the type of guys he spent time with. After all, philosophers spend a *lot* of time wondering. It's kind of their jam. A casual conversation over a few glasses of wine with my uncle Bruce, a philosopher at Wayne State University, can leave one feeling as "wise" as Socrates in a heartbeat! But many of the other people I know are more like Linda—pragmatic and selective about the things that make them curious.

Psychologists who study personality traits for a living have come to similar conclusions about the different ways people wonder. Their research, based largely on self-reports about curiosity, suggests that people actually differ quite a bit in how curious they are "by nature." Although everyone's curiosity levels can wax and wane depending on what they're doing, a phenomenon called a *curiosity state*, there are also more stable differences in curiosity *between people* that can be observed across many different time points and contexts, called *curiosity traits*. To make things slightly more complicated, curiosity traits also come in two different, but related, flavors: differences in the desire to acquire facts, or *epistemic curiosity*, and differences in the desire to experience new things through our senses, or *perceptual curiosity*. This chapter will primarily focus on epistemic curiosity, simply because it is the flavor with the most neuroscientific research to date.

So let's figure out how curious you are by nature. I've borrowed items from a few different curiosity-related measures to help. Like you did in "Mixology," read each statement, and think about how accurate or inaccurate you feel the statement is, *on average,* unless it includes a more specific time qualifier like "right now." I've included the same scale we used in "Mixology" for consistency.

CURIOSITY MEASURE

−3	−2	−1	0	1	2	3
INACCURATE						**ACCURATE**
Strongly	Moderately	Mildly	Neutral	Mildly	Moderately	Strongly

1. New ideas excite my imagination. ____
2. I like to take things apart to "see what makes them tick." ____
3. I enjoy learning about subjects that are unfamiliar. ____
4. Right now, I feel inquisitive. ____
5. New situations capture my attention. ____
6. It excites me to have a new idea that leads to even more ideas. ____
7. I am currently speculating about what is happening. ____
8. I find it interesting to think about contradicting ideas. ____
9. I like to understand how complicated pieces of machinery work. ___
10. I feel involved in what I am doing right now. ____
11. I like solving puzzles or riddles. ____
12. I like to ask questions about things I don't understand. ____

Each of these statements relates to some aspect of curiosity. The more frequently you found yourself agreeing, on average, the more curious you are *in general*. To get more specific, calculate your average level of *epistemic curiosity* by adding your scores for questions 1, 3, 6, and 8 and dividing by 4. Sanity check: The number should be between −3, which corresponds to strongly not curious in this way, to +3, which corresponds to high levels of epistemic curiosity. Now let's calculate your average level of *perceptual curiosity* by adding your scores for questions 2, 5, 9, and 11 and dividing by 4. Once again, the closer your value is to +3, the higher your level of perceptual curios-ity, and vice versa. Finally, questions 4, 7, and 10 are measures of your current, or *state-level of curiosity*.[8] There are only three of them, so divide the sum of your scores by three this time.

So, how curious are you?

Because these are personality measures that are normally distrib-uted, I would expect most people in the real world would score be-tween −1 and +1 on each dimension. However, if you're taking this assessment, it's pretty likely that you're also someone who picked up—and, hopefully, read—almost an entire book about how your brain works. I might be biased, but that doesn't *seem* like the kind of thing that someone who doesn't like to ask questions about how things work would do. But before I get too far ahead of myself, let's talk about the brains of people who are more or less curious by nature. Based on what you've learned so far, do you think there would be a way for me to tell, if I were to scan your brain, whether you might be interested in my book?

Which came first, the curious chicken or the knowledgeable egg?

If you're curious about what a curious brain looks like, you might enjoy reading the case study of an exceptional brain that belonged to a man who claimed, "I have no special talents. I am only passionately

---

8. Measuring your curiosity about curiosity is so meta. And if you went through the mental effort necessary to notice that I didn't categorize question 12, I'll take it as evidence that your state-curiosity level is high! The truth is that question 12 is typically categorized as epistemic curiosity, but I think it's a bit more nuanced depending on the type of question you're asking. If you're asking about how a pendulum works while observing and interacting with it, for instance, you might very well be measuring perceptual curiosity!

curious." The man was Albert Einstein, and the extent to which he had "no special talents" is highly debatable.

Regardless of the accuracy of his self-assessment, Einstein's brain has been photographed and measured posthumously and is thoughtfully described in a series of papers by neuroanthropologist Dean Falk and her collaborators. And as you might guess, it is exceptional in many ways. Among them are the fact that he has notable expansions in his flashy prefrontal cortex, the center of "goal-directed" thinking, in both hemispheres.[9]

But the question is, did these peculiarities *cause* Einstein to be passionately curious, or did they arise because of the massive amounts of knowledge his brain amassed after a lifetime of passionate curiosity?[10] In other words, is Einstein's brain like an exaggerated version of the London cabbies' brains? If so, what might the costs have been? Given that we didn't have the technology to measure his brain longitudinally like Maguire did with the cabbies, there isn't a good way to disentangle these factors.

Unfortunately, the same limitations apply when interpreting the results from the brand-new area of research exploring the neuroscientific basis of individual differences in curiosity. However, results from this nascent line of research do provide certain advantages over studying the brain of a dead guy. For one, these studies actually *measure* trait curiosity in hundreds of living participants. They also use contemporary neuroimaging methods to investigate the properties of these brains at approximately the same time that their curiosity levels are measured. Doing so allows researchers to begin to

---

9. To be more specific, he has *four* gyri instead of three in his frontal lobes. It's wild. His brain is so convoluted it looks like a pile of Easy Cheese on a cracker.

10. Of course, there are a lot of challenges with trying to figure out how the shape of a dead guy's brain relates to how he operated. It's also entirely possible that his brain's structural idiosyncrasies were only tangentially related to his curiosity and intelligence through the kinds of computations they allowed him to perform.

systematically investigate the features of brains that vary in people who are more or less curious "by nature." The results of this research align around one important fact—that individual differences in trait curiosity do not live in any specific part of the brain. Unlike the large size of Einstein's "hand knob" in the right hemisphere, which tells us something about how skilled his left hand was,[11] there is no "curiosity knob" in the brain. Instead, the brains of people who are more or less curious by nature differ in ways that relate to how *in sync* they are.

The doctoral thesis of Ashvanti Valji provides a nice summary of this work. Valji was interested in how both epistemic and perceptual curiosity levels related to the organization of selected high-speed, white-matter pathways in the brain. One of the pathways she zeroed in on was the inferior longitudinal fasciculus, or ILF. The ILF is a massive bundle of white-matter neurons that carries information between the visual areas in the back of the brain and a region in the front of the temporal lobes called (unoriginally) the anterior temporal lobe, or ATL.

Though the functions of the ATL are still debated in the field, many people agree that it forms a hub in the brain where the different bits of information you know about a thing get stitched together. Take, for example, a coffee cup—the vessel that bears the nectar of the gods. As you learned in "Navigate," the representations of objects like this are widely distributed over different regions of the brain. This is because the neurons in your brain that know how to recognize a coffee cup *visually* are physically far away from those that know how to *use* a coffee cup. Where does your hand go, and where does the coffee go? These neurons are also far away from those that program

---

11. Sometimes this fun fact about Einstein's brain is used as evidence that he was left-handed; however, his biographers insist that he wrote with his right hand. He did play the violin, though, and a similar enlargement of the left-handed sensory and motor representations has been observed in skilled violinists by using modern neuroimaging techniques.

the motor movements involved in the different actions you can perform on the cup—pouring coffee into it, reaching for it, holding it in your hand, lifting it to your lips, and so on. And the motor-planning neurons are also in a different part of the brain from the neurons that know what the verbal label for "coffee cup" is. Then, there are still other groups of neurons involved in recognizing the name of the coffee cup if it comes in through your ears, in reading it if it's printed somewhere, or in producing the name for it if you want to speak. That's a lot of knowing about a coffee cup, distributed across many different brain regions!

To measure the organization of the ILF, the information superhighway that carries data from the visual part of the brain to the knowing hub in the ATL, Valji used a technique called *diffusion imaging* in 51 healthy young adults. Diffusion imaging tracks the movement, or diffusion, of water molecules throughout the brain. This information is then used to make inferences about how many white-matter neurons there are in any part of the brain and what direction they're traveling in. In short, because water molecules can't easily move across the fatty insulation that covers white-matter neurons, the molecules that find themselves coexisting with large white-matter tracts are more likely to move parallel to the direction of information transfer than perpendicular to it. And for a variety of reasons, this movement is easier to measure than the direction of the neurons themselves.

The strongest relation Valji observed was between trait levels of epistemic curiosity and the organization of the ILF in both hemispheres. Her results showed that people with higher epistemic-curiosity levels also had lower diffusivity (more restricted water movement) in the ILF than did people with lower levels of epistemic curiosity. This finding might reflect two underlying differences in the ILF: (1) that more-curious people have *more* white-matter neurons traveling between visual regions and the ATL, or (2) that the white-matter neurons that make up their ILF have a more parallel organizational structure. In other words, the brains of people who are less curious "by nature" may simply have more off-ramps on their information superhighways than the brains of more-curious people do. Critically for how you function, either explanation would create greater bandwidth in more-curious individuals, allowing more high-speed information transfer between the cortical processing centers involved in recognizing what you see and those that integrate that information with everything else you know about it.

In other words, the results from this study suggest that people who are more curious by nature also have meaning maps that are more in sync. They also demonstrate that the biological reflections of a person's tendency to explore new idea spaces are not localized to one brain region. Instead, more-curious brains seem to have more coordination between the bits of knowledge distributed throughout the brain. The effect of this would kind of be like stacking the dominoes in their memory maps closer together. Knock over one and you're likely to make even more connections to other ideas.

Unfortunately, this information doesn't help us understand whether greater neural coordination *causes* people to be more curious, or if it's a *consequence* of acquiring a larger database of "fun facts" through greater exploration. One limitation to this body of research is that the scientists were using rather static, or "trait," measures of curiosity, and mapping them to other *relatively*[12] stable indices of brain connectivity. To get better traction on this chicken-and-egg problem, we'll need to explore how wondering (and wandering)

---

12. I use the word "relatively" here because, as you've learned, brain connectivity is dynamic. But unlike the research we're about to talk about, the white-matter measures used by Valji are not sensitive to moment-to-moment changes in curiosity.

shapes the mind and brain in the *moment* that new information is being consumed.

How does curiosity fuel learning?

Catching a curious brain in the act is tricky. Among other things, it requires creating a condition that might pique the curiosity in the first place of a person you don't even know. And don't forget that you've got to do this while they're sitting in a laboratory or lying still on their backs in a noisy MRI tube. But the clever neuroscientists studying curiosity have developed a variety of tasks that seem to do just this.

Min Jeong Kang and colleagues, who first captured curiosity in the MRI-scanning environment, did so using a trivia experiment like I described at the beginning of this chapter. First, trivia questions were presented through a mirror so they could be viewed while lying on your back at the noisy tube. Then, participants read each question and rated both their curiosity and confidence about each answer. After a dramatic pause so that their corresponding patterns of brain activation could be recorded, the answer to each question was provided. When the scientists compared the brain activation recorded when participants read the questions they were *most* curious to learn the answers to against those recorded when they read questions they weren't very interested in, a pattern emerged. Unlike the distributed connectivity effects observed with differences in curiosity between people, changes in moment-to-moment curiosity levels were associated with a few *specific* brain regions. Among them were our old friends the basal ganglia nuclei and their important collaborator, the flashy prefrontal cortex that was enlarged in Einstein.

How might this result help us to disentangle the chicken-and-egg problem of curiosity research? The fact that temporary changes in curiosity are localized to the basal ganglia and prefrontal cortex, while differences in more stable curiosity levels between people are more widespread, suggests that the latter might be *more* related to what one comes to know about the world through their explorations of it than to how their hunger for knowledge drives these explorations.

Evidence consistent with this idea can be found in skill-training studies, which often show that connectivity between cortical regions increases as people learn new skills. For example, the white-matter tracts that make up the ILF, which were related to individual differences in epistemic curiosity in Valji's experiment, have also been shown to *increase* in their bandwidth after young adults spend six days learning Morse code. Imagine how those differences might scale up when you compare the lifetime of decisions that more- or less-curious individuals make to explore and learn about new territories.

So let's return to the basal ganglia for a second and talk about their role in the questioning-and-answering cycle. What business do my favorite brain areas have playing trivia? Because the basal ganglia nuclei are evolutionarily older than the hills, it's likely not what they were made for![13] To understand how the basal ganglia are involved in curiosity, you'll need to become like an ATL and form links between the different pieces of knowledge you've acquired about them throughout this book. First, in "Focus," you learned about how the basal ganglia nuclei impose order on the massive amount of information traveling to the prefrontal cortex. Based on a particular context or goal, the basal ganglia can "turn up" the volume on the signals they deem important, and "turn down" the volume on those deemed less important. If this helps to set up the prefrontal cortex for successful navigation, and the outcome is better than expected, we learned in

---

13. Lizard trivia night would probably be pretty *basic*.

"Mixology" and "Navigate" that dopamine gets released. This, in turn, facilitates the rewiring that helps you learn about, and remember, what you *did* to get to that good thing.[14] But before you can connect these "fun facts" in your knowledge map about the basal ganglia to your curiosity level, you need one last piece of the puzzle—a description of how (and when) dopamine signaling helps us to navigate back to those good things.

When we first started talking about dopamine in the "Mixology" chapter, I described a hypothetical situation in which a random walk through a new neighborhood landed me squarely in front of an ice-cream shop. I think it's safe to say, though, that unless you're a lot luckier than I am, this is not the way we do things in real life. Sure, the adventurous, extraverted, perceptually curious among you probably go on walks to explore new places. But unless you're on a "coin toss" walk, I doubt your choices about which way to go are really random.[15] Instead, during your explorations, you likely choose a neighborhood that you have some reason to be interested in in the first place. Then at each corner, or decision point, your brain would likely use *clues* to guide your decisions about which way to turn. "I see a bunch of trees off to the left, and I feel like looking at nature, so let's go that way," your horse navigation might suggest. Or—depending on your mood—"I hear traffic to the right, and I feel like finding shops, food, or other signs of civilization. Let's head that direction!" Of course, if you're more of a stick learner, your brain might sound more like "I hear traffic to the right, and I prefer quiet, let's go left" or "I see a bunch of trees to the left, but I prefer civilization—let's go right!" In either scenario, it uses the data in front of it to drive its decision about what to do next.

---

14. And of course, when bad things occur, the reverse happens.

15. A coin toss walk is one in which you flip a coin to decide whether to go left or right at each corner. And even if you did take such a truly random walk, you would likely start by choosing a neighborhood you felt safe in!

Our third "fun fact" about the basal ganglia explains how such navigation is accomplished. As it turns out, feel-good dopamine signaling doesn't *just* occur when good things happen. Instead, your basal ganglia use strategically released bursts of dopamine, coupled with everything they've learned about outcomes of actions, to create "warmer" or "colder" suggestions that draw you toward the good things in life (or away from less-good ones). In the lab, if you present a cue, like a light or a tone, right before a rat gets a food reward, you will start to see dopamine neurons respond more and more strongly to the cue. Eventually, *most* of the feel-good dopamine will be released at the cue, as opposed to when the food comes. In other words, their brains celebrate success at the earliest point at which they can be confident that a reward is coming. This is why clicker training works for pets. When the cue is a strong predictor that a reward is coming, it becomes rewarding in and of itself.

So what does this have to do with trivia? When you stitch these three pieces of basal ganglia knowledge together, a picture of dopamine as "The Wick in the Candle of Learning"[16] emerges. The curiosity that people *feel* is a sign that their brains have computed that there is a high likelihood the information discovery process will be rewarding. Or, for the stick learners, they have decided that there is a low likelihood that their information-finding experience will turn out worse than expected. And if these people were out exploring in the real world, their brains would strategically release bursts of dopamine to help them navigate down the path that was most likely to lead them to a reward, whether it be a piece of information or an ice-cream shop.

In Min Jeong Kang's trivia experiment, however, no exploration was necessary. The answer to every question was provided shortly

---

16. This is the catchy title of Min Jeong Kang's paper, and I'm a sucker for a catchy title!

after participants rated how curious they were. In this unnatural laboratory environment, all participants needed to do to get their reward was to wait. While they did, their basal ganglia released dopamine in celebration of the knowledge reward they knew was coming. And as you might suspect, this dopamine burst facilitated rewiring in their brains. As a result, they were better able to remember the answers to the questions they were most eagerly awaiting.

But what would happen in the real world when you don't know for sure whether a particular action would lead you to find rewarding information? What about the potential risks that might be associated with exploring the unknown? We'll dig into these questions in the next section, with a more careful consideration of the costs and benefits associated with discovering the unknown.

Curiosity during uncertainty

When the results of Kang's groundbreaking curiosity experiment came to light, many researchers wondered how the relationship between curiosity and learning might work in the real world. To get a better handle on how the basal ganglia and prefrontal cortex would react in more complicated learning situations, they each changed the trivia paradigm in different ways. Romain Ligneul and collaborators were interested in the element of surprise. Their experiment, like many others, was also centered on trivia questions. But this time, all of the questions were about movies!

The neuroimaging part of their experiment began much like Kang's. Study participants read trivia questions while lying in the scanner and indicated how curious they were about the answer. However, this time, like in real life, there was no guarantee that the answers would be provided. Instead, the first half of their experiment worked like a coin-toss trivia game, in which answers were randomly presented after 50 percent of the questions. This was the "high sur-prise" condition, which was compared to a subsequent "low surprise" condition in which all answers were provided.

So what does a curious brain look like when it can't be sure whether it'll be rewarded with an answer? Results from Ligneul and colleagues showed that activation in the prefrontal cortex increased, but the basal ganglia remained silent. It's too early to celebrate, they decided. Instead, it was only when the answer was presented that basal ganglia activation increased. How do you think this might in-fluence what participants learned?

When memory was tested in a post-trivia pop quiz, participants' ability to remember the answers to movie trivia questions was influ-enced both by how curious they were about a question and by how surprised they were to receive an answer.[17] This plot twist provided another piece of evidence linking dopamine and the basal ganglia to learning. Participants remembered the answers presented in the first block, or "high surprise" condition, better than they remembered the answers presented in the second block, or "low surprise" condition, regardless of how curious they were about any specific answer. In fact, only when the answers were presented after each question did their results replicate Kang's, with the level of curiosity about an an-swer predicting the later memory of it. When getting any answer was uncertain, and the basal ganglia remained silent, participants' curios-ity levels didn't significantly predict how much they would learn. Remarkably, they were more likely to remember the answers they weren't curious about at all, when they were presented in a "high sur-prise" condition, than they were to remember the answers they were most curious about, when presented under conditions of certainty.

Why might *surprise* trump *interest* when it comes to learning

---

17. To measure the effect of surprise, they compared activation in this 50 percent answer block to a second block of trivia in which the answer to every question was given.

things? The answer (of course) lies in the mechanisms of your basal-ganglia-driven reward systems. As you may recall from our discussion of how it might not be as good as you'd expect to be Beyoncé, the basal ganglia quickly adapt to good things when they're expected. What really gets them going is an *unexpectedly* good event. This can mean finding something even the slightest bit good when you weren't expecting anything, or that something turns out better (or even less bad) than you expected it to be.[18] From your basal ganglia's perspective, surprises are life's *most* "teachable moments," whether you learn what to do (carrot) or not to do (stick) in the future.

The results from this experiment provide additional evidence that our human brains consider knowledge to be rewarding. The more curious a person is to learn the answer to a question, the bigger they anticipate the knowledge reward to be. And when they receive "surprise" knowledge, their brains respond like they stumbled upon an ice-cream shop. Both situations involve the release of dopamine at the earliest time that information can reliably be expected, which promotes the rewiring required to learn.

Another experiment, conducted by the coauthors of the PACE framework, Gruber and Ranganath and their collaborators, further explored the influence of curiosity on learning with another clever manipulation to the standard trivia paradigm. Their experiment began in the traditional way, with participants reading trivia questions and providing curiosity ratings. But then, right in the middle of the anticipatory delay between the question and answer, the experimenters inserted a picture of a face. As a clever ploy to make sure people would pay at least a *little bit* of attention to it, the experimenters asked participants to indicate with a button press whether they thought the

---

18. Remember that this is particularly true for extraverts, which makes me want to know more about how extraversion might interact with curiosity-driven decision making!

person in the picture knew the answer to the question.[19] Ninety percent of the time, the answer to the question was then presented. But to keep participants on their toes, the other 10 percent of the time, faces were followed by a string of X's, implying that the person didn't know the answer.

Note that this experiment also involved some degree of uncertainty. But the results of Gruber and Ranganath's study suggest that getting an answer 9 out of 10 times was a good enough reason for their participants' basal ganglia to celebrate early. As in Kang's original experiment, Gruber and Ranganath found that curiosity was associated with increases in both basal ganglia and prefrontal cortex activation during the *question* phase of the experiment. And as predicted, they also found that people were more likely to remember the answers they were most interested in when given the post-trivia pop quiz.

But what about the faces? What happened when a brain, anticipating an information reward, saw a face along the way? One of the most novel contributions of Gruber and colleagues' experiment was to see whether participants also learned the "incidental"[20] information presented during curious moments better. To do so, they gave them another "pop quiz" in which they were asked to identify which faces they had seen previously from a series that contained both new faces and those used in the experiment.

According to their results, people also recognized the faces that were presented between the questions and answers they were most

---

19. To the best of my knowledge, these answers were not analyzed. In the real world, however, you might expect people to pay more attention to the face of someone they believed knew the answer than to one they didn't. I also have no idea what the characteristics of these faces were, and what kind of implicit biases people may have used to judge—based on someone's face—what they knew.

20. We should put the word "incidental" in scare quotes here because in the real world, remembering the face of a person who could potentially answer your question would not be incidental at all!

curious about *more frequently* than they did the faces that were presented before an answer they didn't care as much about. This finding suggests that the early dopamine release associated with the anticipation of an information reward opened a window for learning. Through that window, some additional information crept in and received a memory boost!

But before you clever parents or educators start trying to use this data to concoct brilliant or diabolical ways to teach boring things, I should mention that the *effect* of curiosity on incidental face learning was much smaller than it was on learning the trivia answers—the information deemed rewarding in the first place. Although the increase was statistically significant, recognition of faces presented between high-curiosity questions and answers increased only 4.2 percent over those presented between low-curiosity questions and answers at the group level. By comparison, memory for the answers to high-curiosity questions was 16.5 percent higher on average than memory for the answers to low-curiosity questions was. This difference in curiosity-driven learning may be explained by signal routing in our basal ganglia control mechanisms. Because the basal ganglia could learn to predict that irrelevant, interfering face stimuli would come before the answers they were seeking, they might learn to turn down the signals coming from the irrelevant face stimuli.

To explore the possibility that the curiosity-induced-learning windows may open at different times, or to different degrees, in different participants, Gruber and colleagues measured changes in brain activation in high- versus low-curiosity trials, in the early delay period between the questions and the faces. As you might expect based on everything we've been discussing, there were *huge* individual differences in both the *degree* and *direction* of brain activation changes. Only about half of the participants showed changes in the direction predicted by the group average—that basal ganglia activation *increased* when people were curious about the answers they were waiting for. The rest of the participants showed little to no increase or even a *decrease* in basal ganglia activation before the faces were presented. This is consistent with the idea that the basal ganglia in these participants were "turning down" the signal for the irrelevant faces. Not surprisingly, the people whose basal ganglia decided to celebrate early also showed the biggest incidental learning effects for the faces. Several even showed improvements in the 10 to 15 percent range, in line with those observed for the answers.

When we connect the dots on this body of research on how curiosity shapes learning, a consistent pattern emerges: The basic reinforcement-learning mechanisms that our brains evolved to drive us toward good things also motivate us to seek knowledge rewards. The strength of the learning that occurs during our explorations relates both to how rewarding an individual brain perceives a piece of knowledge to be and to how likely they think they are to receive any information at all. So whether you're on a walk in a new neighborhood, playing a game of trivia, or browsing the shelves at your favorite bookstore, the curiosity you experience is your brain's way of strategically dropping bits of *feel-good* dopamine rewards that guide you in the direction it believes will lead to the best information. But what if it's an uphill battle (literally or figuratively) to get that information? In the next section, we'll explore exactly how strong the siren call of dopamine reward signaling is by looking at the *cost* curious people are willing to pay to get a piece of information.

The costs of curiosity: How bad do you *really* want to know?

In May 2020, Johnny Lau and collaborators published what is arguably the most "metal" curiosity experiment of all time. In doing so, they illuminated the dark side of what it means to be curious by nature. "Curiosity is often portrayed as a desirable feature," Lau writes in his abstract. "However, curiosity may come at a cost that sometimes puts people in harmful situations."

In the real world, the potential costs of curiosity are wide-ranging. At the most vanilla end of the spectrum, you have costs like the amount of time you spend finding information. For those of us who are prone to go way, *way* down a rabbit hole in our information exploration, this might mean dozens of hours spent reading about alpaca, or black holes, or other bits of knowledge we are unlikely to use in our daily lives. Then there are the trickier, and more dangerous for many, social costs. Your willingness to ask a question in front of others, for instance, is likely influenced both by how curious you are about the answer and by how worried you are about being publicly embarrassed. Despite hearing "there's no such thing as a stupid question" throughout our lives, we all know that there is ample opportunity to ask an embarrassing one. Then, at the even riskier end, curiosity might drive someone to experiment with drugs or engage in any number of other thrill-seeking behaviors.

To understand *how motivated* people really are by their curiosity, Lau and colleagues measured what price they were willing to pay in exchange for information. The experimenters also did something clever to ground the motivation for knowledge in a more relatable, real-world context. They made their participants hungry! People were asked not to eat for several hours before the neuroimaging study. This way, *food rewards* could be used as the baseline against which *information rewards* were compared. How does a brain hungry for a hamburger compare to one that's hungry for knowledge?

When the hangry participants entered the scanner, they saw one of three types of trials. The first type was a trivia question like the other studies we discussed: Read a question. Rate how confident you are that you know the answer. Indicate how curious you are. The second type of trial was different in a pretty exciting way—instead of reading trivia, participants saw video clips of magic tricks! What followed was in the same spirit as the trivia questions: How confident are you that you know how the trick was accomplished? How curious are you to know how the trick was accomplished? In the third condition, people saw pictures of different types of food rewards. When they did, they were asked to indicate how much they would like to eat that piece of food. When interpreting the results of this study, it's important to keep in mind that this was not an abstract, "How much do you like hamburgers?" question. Instead, they were asking hungry participants whether they *wanted* a hamburger or not. And much like there was a real possibility of receiving a knowledge reward following trivia questions and magic tricks, participants were told that there was a real possibility of receiving the food depicted in the picture at the end of the experiment.

And here's where things get *really* interesting. After rating their desire for either knowledge or food, study participants were given the opportunity to put their money where their mouths were, so to speak. Specifically, the strength of a person's desire was measured based on the price they were willing to pay to receive the object of that desire. After each trial, they were given a choice—risk getting an electric shock for the possibility of the reward presented in that trial, or pass on the opportunity. The likelihood of receiving a shock versus a reward (food or knowledge) ranged from a 16.7 percent chance of being shocked (1 out of 6 times) to an 83.3 percent chance of being shocked (5 out of 6 times).

After viewing a trivia question, a magic trick, or a piece of food, participants saw a pie chart that visually indicated the level of risk of shock versus reward on the current trial, and they were asked to decide whether they wanted to take the gamble. To help you understand the participants' state of mind, it's worth mentioning that before they got into the scanner, they were given electrical shocks of different strengths to determine their pain thresholds. The goal was to deliver a shock at a level that would feel *uncomfortable* but not extremely painful.[21] The reason I think this bit of information is important is that the fact they actually experienced shocks of different intensities must have grounded the participants' decisions in the real-world, visceral experiences of what they were risking.

As you might expect, for most participants, when the likelihood of getting shocked *increased*, their overall willingness to take the gamble *decreased*.[22] However, as their self-reported *desire* for the reward increased, so did their willingness to risk shock to gain rewards. Importantly, the overall pattern of responding looked very similar for food rewards and for information rewards. This provides pretty strong support for the idea that our desire to know things arises as part of an evolutionarily older reinforcement learning system.

And now for the million-dollar question: What is going on in the brain of someone who is willing to risk getting a shock to learn how a magic trick works or what the answer to a trivia question is? To answer this, Lau and colleagues explored patterns of brain activation for trials where participants eventually took the gamble to those where they didn't, over two critical points. What they found was consistent with the idea of dopamine release in the basal ganglia working like a game of "warmer/colder" to motivate participants to seek information rewards, even in the face of real costs. Small increases in basal ganglia activation were observed during the initial viewing of the items that would eventually lead people to take risks. This could be really good, their basal ganglia reported—consistent with the idea that their brains were evaluating how rewarding a decision to explore was likely to be. But much larger and more widespread differences in

---

21. I told you this study was metal!

22. It's worth noting that two participants were excluded from the study because they accepted *every* gamble. This just goes to show that some participants are also really metal.

basal ganglia activation were observed during the actual decision phase. When the real-world costs were in front of them, and it was time to decide how to *behave*, the siren call of the basal ganglia was strongest.

But wait, there's more! In a follow-up exploratory analysis, the research team decided to ask the critical question "Who are the basal ganglia talking to?" by measuring patterns of synchronization between the basal ganglia and all other brain regions at the point of decision. What they found was pretty startling. When participants decided to risk getting shocked, there was a significant *decrease* in connectivity between the basal ganglia and the parts of the sensorimotor cortex that have been associated with virtual feelings, or anticipation, of pain.

Here's what this pattern of results suggests to me, based on my understanding of basal ganglia signal-routing mechanisms. When the basal ganglia estimate (based on your previous experiences and their understanding of the goal at hand) that a piece of information is worth taking a risk for, they actually *turn down* the signals coming from a part of the brain (in this case the pain anticipators) that might influence the prefrontal cortex to make a different decision! It seems pretty straightforward to generalize from these results what might be happening in the brain of someone who *decides* to free-climb El Capitan, or to engage in other risky behaviors, despite *knowing* about the risks.

The results provided by Lau and colleagues provide a rich context for thinking about curiosity neuroscience with a real-world perspective. Not only are curious people driven to explore the unknown, they also take real, calculated risks in doing so. In fact, according to the "A" in the PACE framework, there is an *appraisal* phase that brains go through in the wild *before* they feel curious. Perhaps this is prudent, as Lau's results show that if given the opportunity to feel curious before knowing what the potential risks of a situation are, our curiosity can "turn down" our brain's ways of signaling potential dangers.

In the last chapter of this book, we're going to talk about one of the riskiest, but most essential, human opportunities to explore an unknown territory we can never observe directly—the mind of another. But before we do, I'd like to rehash what we've learned about the cycle of wondering and knowing, with a focus on the costs and benefits of our willingness to explore new spaces.

Summary: When facing the unknown, our brains decide whether to *explore* or whether to *ignore* based on the estimated value of knowledge

Many of the findings discussed in the last two chapters of this book fall into place when we view them according to the PACE framework outlined by Gruber and Ranganath. Why are we motivated to learn new things, even if those things might not directly relate to something we may do in the future? As you learned in "Navigate," each new piece of information we acquire shifts our knowledge maps around a bit, because this knowledge becomes connected to other things we know. And at the center of this knowledge map is our understanding of ourselves and our place in the universe. So even though I might *never* go into outer space, or need to estimate the age of a jellyfish, the things I learn about infinity or immortality change the way I think about myself. Importantly for the framework, they also change the kinds of *predictions* I might make about what I might find in unexplored spaces.

But the fact that *you* lie at the center of your meaning map also creates a risk that we have not yet discussed: What if acquiring a new piece of information has the potential to change the way you understand the world in a way that feels threatening to your identity?

The answer, I *believe*, can be partially explained by the *appraisal* phase in the PACE framework. According to the theory, it's only when things are deemed *relatively safe* that your brain creates the feeling of *curiosity* that motivates *exploration*, the last two phases in the cycle of wondering and knowing. Of course, this is a huge opportunity for individual differences to crop up, because the level of risk that any individual finds acceptable must be tied to their previous experiences with information-seeking, as well as to whether their brains learn more by carrot or by stick. But it's also important to consider how differences in the "space" you're exploring, physically or metaphorically, may feel more or less threatening.

Much like physical threats, such as the risk of receiving a shock, might shut down our willingness to explore, your brain must also be motivated to protect itself from psychological threats. If this is true, maintaining your most central, identity-based *beliefs* must be one of the goals your prefrontal cortex would use to guide your behaviors. The power of this kind of belief structure is that rather than driving you to explore, or collect statistics and form an objective opinion about what might be true of the world "out there," such top-down navigation strategies would cause your basal ganglia to turn up the volume only on "relevant" information—that is, information that is consistent with your identity-based beliefs—while turning down the volume on any information deemed "irrelevant" because it doesn't support your worldview. In a recent opinion paper, Jay Van Bavel and Andrea Pereira outlined how such a brain-based model might be used to describe the relation between personal values, political beliefs, and partisan behaviors.

Whether you're willing to *believe* it or not, we *all* do this. It keeps us feeling safe, and protected, and correct. Psychologists who study how we form and hold beliefs have long known that when people are surprised by information that is inconsistent with what they *believe* to be true, they don't often behave rationally. Instead, they ignore or even discredit evidence that is inconsistent with their beliefs—a phenomenon known as a *confirmation bias*. The important take-away from this is that the possibility of encountering a piece of information that is inconsistent with your centrally held beliefs may be perceived as a threat during the *appraisal* phase. The result would be to shut down the cycle of wondering—and wandering—that brings us to explore the unknown. With this in mind, in the next chapter, I will walk you through one of the most vulnerability-inducing explorations of the unknown that any of us can ever undertake—our attempt to see through the bubbles created by our own brains to connect with another person, whose views of the world may not align with our own.

CHAPTER 8

CONNECT

How Two Brains Get on the Same Wavelength

In the book *Talking to Strangers*, Malcolm Gladwell takes his readers through a series of dramatic, real-world examples of misunderstandings between people that make two points very clearly: (1) Understanding other people can be *really* difficult; and (2) misunderstandings between people can have disastrous consequences. From Ponzi schemes to genocide, when we get it wrong—we *really* get it wrong. Is it any wonder that some of us are hesitant to explore relationships with others?

In this book, I've provided the background knowledge to help you understand the biological barrier that can drive misalignment between people. When two different brains, shaped by the confluence of their unique biology and life experiences, interact with each other in a *shared* environment, they do so through the barrier of the different *subjective* realities that they create.

Yet the very brains that make it difficult to see eye to eye with someone else are the same ones that inspire us to try. And though it's certainly truer for some more than others, our social human brains *crave* connection. From early caregiving relationships through the different types of intimate adult partnerships we form, our brains contain a host of built-in mechanisms that drive us to connect. And this makes good sense, given how central relationships are to our survival. In fact, forming connections with others is one of the most important brain functions of all.

I think George R. R. Martin got to the heart of this when he wrote, "When the snow falls and the white winds blow, the lone wolf dies but the pack survives"—as a central part of the House Stark narrative in *Game of Thrones*. Because when times get tough, having close relationships is critical to our survival, even if most humans no longer rely directly on one another for things like warmth or hunting in packs. We have seen this play out over and over in health-related research, where the power of touch has been shown to help the brains and bodies of premature babies develop, and social support networks help buffer the health effects of chronic illnesses like AIDS. To put the importance of close relationships into a more concrete, health-related context, consider the results of a recent meta-analysis conducted by Julianne Holt-Lunstad and Timothy Smith. After analyzing data collected from more than 300,000 participants around the world, the authors concluded that lacking close interpersonal relationships was more than twice as strongly associated with early mortality as either excessive drinking or obesity.[1]

And I'm sure that our understanding of the health risks associated with loneliness will increase exponentially as psychologists and healthcare professionals start to crunch the data collected from the massive "social isolation experiment" that the COVID-19 pandemic created. A quick search through the scientific literature using keywords "social isolation," "pandemic," and "health" returned more

---

1. Excessive drinking was defined as greater than six drinks per day and was compared to abstinence. Though they did not specify the cutoff, obesity was defined based on the body mass index of obese versus lean. They also showed that a lack of close interpersonal relationships was about equal to smoking up to fifteen cigarettes per day!

than 1,500 articles published in the last two years on this topic. Though I can't *quite* get to the place of calling this a silver lining, each of these studies will contribute to our understanding of *why* interpersonal connection is a necessary component of healthy living, and which of the many ingredients of successful connections promote physical and mental health. But will they provide the recipe for forming healthy relationships?

Fortunately, my colleague Jonathan Kanter, director of the Center for the Science of Social Connection, has some help for us in this area. In 2020, he published a model defining the coachable ingredients of intimate interpersonal relationships, which includes three types of bidirectional information exchange between minds and brains: (1) nonverbal emotional communication, in which the expresser of the emotions feels *safe* being vulnerable; (2) verbal expression of the self, in which the expresser feels *understood and validated*; and (3) asking or requesting behaviors, in which the expresser feels *helped*. At its heart, based on a body of previous research on relationships, Kanter's model defines the conditions that bring people closer together. In doing so, he also describes some of the places where misalignment can drive us apart.

According to Kanter and the authors of his model's predecessor, when interpersonal exchanges meet the criteria he outlined, the result is rewarding. It *reinforces* the desire to connect and increases the strength of the bond between the pair. But when they don't, the opposite occurs. And this makes perfect sense from your brain's perspective, based on the way our feel-good dopamine reward circuits learn based on the outcomes of your actions. Of course, we also talked about the carrot-and-stick learning mechanisms that work in concert to drive us toward good things and away from things that didn't pan out so well in the past. When you consider these different types of learning in light of Kanter's theory on intimacy, you can start to imagine why some people might be more strongly influenced by things that went wrong in previous relationships than others might. In this chapter, we'll build on this idea by adding into the mix another neurochemical that can motivate us to engage in these vulnerable relationships—*oxytocin*.

Kanter's model also acknowledges that bidirectional communication gets processed through each person's "perceptual filter." And you should feel particularly well qualified to think about the complexities that this adds after what you've learned in this book. But as we'll discuss in this final chapter, there are multiple ways to understand the mind of another. Some of them are more automatic, allowing you to put yourself in the shoes of the person you're trying to connect with. But others that require more mental effort may be less susceptible to the kinds of misalignment that happen when two different brains try to communicate.

And as we enter this last chapter of our adventure together, I sincerely hope your increased understanding of the ways our brains drive us through life in different ways might motivate you to *try* to reach across the void and connect with someone else whose brain may have a different perspective on reality from your own. After all, some of the most important collaborations in history happened when people with well-documented *differences* in their ways of thinking, feeling, and communicating came together—like John Lennon and Paul McCartney, Pauli Murray and Eleanor Roosevelt, Bill Gates and Paul Allen, and Susan B. Anthony and Elizabeth Cady Stanton, to name a few. In the remaining pages of this book, we'll discuss both how our brains can drive these meaningful connections and how they can get in the way.

What does it mean to know someone?

Before we start talking about the mechanisms that brains use to connect with others, I'd like to try to define the challenge of understanding another human, from your brain's perspective. In "Navigate" and "Explore," we spent a decent amount of time laying the foundation for what it means to make sense of things, and how we use this knowledge to drive our decisions. So how might knowing another *person* be similar to, or different from, the way we come to understand other phenomena in the world "out there"?

The short answer is that from your brain's perspective, knowing a person is not fundamentally different from knowing how anything else in the world works—except, of course, that humans are *much* more complicated, and harder to predict, than most other things we try to understand. This creates a challenge because your brain *likes* to be able to predict things. As we discussed in "Explore," it's one of the basic tools it uses to figure out whether it needs to get more information or not.

Unfortunately, since humans are capable of such a wide repertoire of behaviors, we *never* have enough information about any one individual to perfectly predict what they're going to say or do next.[2] If we *did*, it might look something like one of my favorite scenes from *Westworld*, in which one of the central characters, Maeve, learns that she's a robot.[3] "No one knows what I'm thinking," she whispers to the empathetic technician, who pulls out a tablet and syncs it to her internal programming. She then proceeds to watch, with alternating anger and confusion, as the words she speaks appear on the screen in what's labeled a "dialogue tree," the instant before she says them. "This can't possibly . . ." she says and reads simultaneously, followed

---

2. Although Andrea certainly *thinks* he can do this—and it can be very frustrating—unless we are talking about science, he tends to be *way off* when he tries to complete my sentences. To be completely fair, these kinds of predictive mechanisms allow him to communicate really well in three languages. But this doesn't make it less *annoying*.

3. I'm referring to the TV show that was released on HBO in 2016, not the film released in 1973, though that was also epic! You can find the scene by searching for "Maeve" and "No one knows what I'm thinking" on YouTube.

by the word "CONFLICT," which pops up in a red bubble labeled "inference engine."

Part of what makes this scene—and the premise on which the show is based—so powerful is that we empathize with the robot. In doing so, we are brought face-to-face with the idea that we are also predictable machines. But one of the big differences between ourselves and robots is the degree of *flexibility* of our behaviors, which we discussed in "In Sync," "Adapt," and "Navigate." As you've learned throughout this book, we don't all respond the same way to the same external situation. And even the *same person* can respond in multiple ways to the same external trigger, depending on the interaction between what's going on in their inner and outer worlds. This is all just to flesh out the point we started with—that understanding others is difficult!

But we *need* to do it. Though things have certainly changed during the pandemic, under normal circumstances we interact on a daily basis with people we know very little about. From collaborating with coworkers to the casual contact that happens when you walk by someone on a busy street and decide how to respond, our brains are called upon all the time to predict how others will behave. To do this, they greedily soak up statistics about human behavior.

You might think of the statistics we collect about people as belonging to a series of concentric circles. In the center circle are statistics about an individual, like "Andrea loves bad jokes," and somewhere near the outside circle are your statistics about humankind, like "uses language to communicate." The circles in between contain information along a variety of axes of specificity that your brain might find useful for navigating any given situation, allowing you to understand what is happening now and what is most likely to happen next. And this is how our implicit biases based on a person's race, age, gender, sexual orientation, political affiliation, socioeconomic status, career, accent, hairstyle, and so on come to influence the way you understand them. As we discussed in "Adapt," some of these biases can have incredibly serious consequences—like associating a person's *race* with the likelihood that the object in their hand is a tool or a weapon. But even those that don't have potentially lethal consequences can really get in the way of seeing someone for who they are, not who you expect them to be. Because the same human brains that are driven to connect with others are also unrelenting pattern detectors, designed to draw conclusions with insufficient data.

Of course, the best way to avoid generalization would be to have a lot of data about each specific person you're trying to interact with. At the end of the day, that advice would lead us to avoid talking to strangers whenever possible. But this takes us back to the ideas discussed in the last chapter. When might we want, or need, to take the risks involved with talking to a stranger? Put a pin in that idea—we'll return to it in a bit. But the idea of having a lot of statistics about one person also raises the question: Is having a lot of *data* the same thing as knowing a person? According to an article written by philosopher Mark White, not really.

And I tend to agree.

For instance, I have collected quite a bit of data this past year and a half about a man who lives in my neighborhood and walks a very old wolfhound mix past my house every day. I can predict certain things about him with pretty high accuracy. Almost every day, rain or shine, he walks his dog down my side of the street, heading southbound, between ten and ten thirty in the morning. The dog is always lumbering along a few feet behind him, as if to say, "You're going too fast, again!" And when my dogs invariably bark their heads off to alert us of the "intruders," the man almost always looks back at his dog. Perhaps because I don't know *why* he does this, the mere fact that I can predict that he *will* do it doesn't make me feel like I know him.[4]

---

4. Sometimes I wonder whether he's looking back to see if his gentle giant is

White's article describes the distinction between knowing someone and knowing *about* them, using the arguments laid out in a philosophical article on the topic written by David Matheson. In his article, White illustrates the difference between *impersonal knowledge*, like all of the strange things we may come to know about a celebrity, and *personal knowledge*, which can only be obtained by interacting with someone directly. But I would add a bit to this argument by saying that sometimes we do *feel* like we know celebrities, or at least I do,[5] while other times a person we spend every day with might feel like a total mystery to us.

The reason for this, I believe, is that our feeling of knowing a person is based on how richly and accurately we think we understand the contents of their mind. While it's clear that having a lot of personal experience with an individual can provide more data, other factors, like how expressive or transparent a person is, must also play a role. Because no matter how much direct experience we have with a person, we can never *see* what's happening inside their heads, in that private and complex inner world that we have spent most of this book talking about. So one of the ways our brains deal with the missing puzzle piece involves the same processes they use to plan and reason about other things they can't observe—by forming an imaginary mental model of it.

This creates a *massive* reverse-engineering problem. Our inputs are the observable data. What did the person say or do, and how did they look when they said or did it? Or perhaps—depending on the type of statistics your brain prefers to focus on—what *didn't* they say

---

somehow provoking my dogs by giving them dirty looks. And other times, I imagine that he's just checking to make sure my ferocious, eight-pound mutt hasn't finally succeeded in killing his dog with its mind.

5. Did you think I was going to talk about Jason Momoa again? Come on—I am not *that* predictable.

or do, and how did they look when they didn't say or do it? In our attempts to model someone else's mind, we have to work backward from the things we *can* observe to try to figure out *why* they behaved the way they did. In the next section, I'll give you some examples of a test frequently used to figure out how good you are at reverse-engineering the minds of others.

How good are you at reverse-engineering a mind?

To figure out how good you are at guessing what someone is thinking based on the clues available to you, let me give you a few items from one of the most frequently used measures of mind-modeling ability in adults—the *Reading the Mind in the Eyes Test* (I'll call it the Eyes Test for short), developed by Simon Baron-Cohen and colleagues. The task is relatively straightforward, but not necessarily easy. Your goal is to try to figure out how someone is feeling based on the expression of their eyes. You know, the whole "The eyes are the window to the soul" thing?[6] For each image on the next page, pick which of the four words surrounding the eyes best matches the feeling they express. Feel free to look up the words in a dictionary if you're unclear about any of their meanings—it's not supposed to be a vocabulary test!

Which of the four mental states do you think best characterizes the person[7] depicted?

---

6. Apparently there's an active controversy about who said this first—Cicero? Shakespeare? Well, we know it wasn't Baron-Cohen, but based on what we've learned using this test, I think it's fair to say that there is important information to be gleaned about a person from their eyes, and it's quite plausible that more than one human in history noticed it.

7. This is definitely Nicolas Cage, isn't it?

irritated   sarcastic

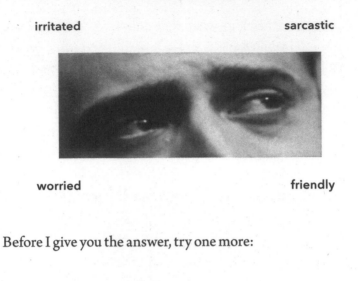

worried   friendly

Before I give you the answer, try one more:

decisive   amused

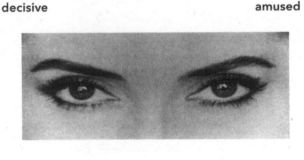

aghast   bored

The correct answers to these items are *worried* and *decisive* respectively. If you'd like more evidence about how good *you* are at reverse-engineering someone based on the expression of their eyes, you can find the whole test online by searching "Mind in the Eyes Test" or by going to the Research tab on my website.

As you'll learn in this chapter, there are multiple ways our brains use the observable behaviors of others to make inferences about their mental states. And this, my friends, is where the proverbial shit hits the proverbial fan when it comes to connecting with others. Because humans are not robots, and modeling our minds in the real world is *not always* as straightforward as a multiple-choice test might make it seem. In the next section, we'll start to describe the parallel paths our brains use to understand others, and under what conditions they can fuel connection or drive us apart.

Reading the mind through a mirror

When it comes to understanding others, the first tool in your brain's tool belt is one that we share with other social primates. It allows us to learn from one another from a very early age, by modeling others' behaviors. Put simply, when your brain watches someone else performing an action, it simulates the way *you* would do that action. This simulation involves "mirror neurons"—a term used to describe groups of neurons that become activated both when *you* perform an action and when you watch someone *else* performing the same action. Through these mirror neurons, your understanding of the behaviors of others becomes coupled with an internal representation of how *you* would perform that behavior—much the same way that your understanding of a coffee cup becomes coupled with the way you would hold the cup.

One strength of this type of mind modeling is that it connects you to others in an organic way that allows you to empathize, or "feel with" them. In other words, this type of mind modeling places you in the shoes of the person you are trying to understand.

This leads us back to what we discussed about meaning maps in "Navigate"—the *you* at the center of the map of your brain's understanding of the world. When we try to understand others, our *default* is to map them to our own, egocentric perspectives. How would *I* be thinking and feeling if I behaved that way? And much like the implicit biases that can shape your assumptions about strangers, this self-projection can be so fast and automatic that you aren't even consciously aware it's happening.

This automatic mirroring probably explains, at least to some extent, a phenomenon social neuroscientists have been documenting over the past five or so years—the fact that we tend to hang out with people whose brains work like ours do. For instance, one clever experiment by Carolyn Parkinson and collaborators started by creating a social network based on the self-reports of 279 students enrolled in the same graduate program. Students who mutually listed one another as friends were connected by a link, and those who didn't were not. If two people were not friends, but listed a mutual friend in common, they were connected in the network through this intermediate link. After creating a network that included all mutual relationships reported by the group of 279 students,[8] the scientists selected 42 of them, with varying degrees of social connection, to participate in a neuroimaging experiment. In it, their brain activity was recorded while they passively viewed a series of video clips, ranging from comedy shows to debates. Afterward, the researchers extracted activation time courses from 80 different brain regions in each participant, and correlated these time series, on a region-by-region basis, for each of the 861 possible pairs of participants.

The results of the study were striking. People who were friends had more similar brain responses across the board than those who only shared a friend in common, who in turn had more similar brain responses than those who did not share a friend in common, and so on. In fact, when the authors used brain similarity to *predict* who would be friends with whom, it explained significant variability even

8. There is one person in this network with no connections, and it makes me feel sad. I just hope this person prefers to hang out with nonacademic peers rather than their stressed-out cohort! ☹

when known predictors like age, gender, and nationality were controlled for. And though this pattern was observed across many brain regions, several of the areas that were most strongly correlated among friends were in the basal ganglia. This result likely reflects something you can probably guess on your own—that people who like the same things tend to like one another better than people who like different things. What this research added to that idea is the idea that people who like one another also have *brains* that respond in similar ways to external stimuli.

But more recent research conducted by some of the same scientists suggests that the magnetic effect of having similar brain functioning extends beyond just responding in the same way to the outer world. For example, Ryan Hyon and team, including Carolyn Parkinson (again!), used the social-network design to study relationships among an entire village—798 individuals, to be exact—on a small South Korean island. This time, the experimenters analyzed task-free functional MRI data from 64 people in the village, with varying degrees of social connection. Again the results showed that the degree of similarity in brain functioning between any two people predicted how likely they were to be friends. But this time, the similarities were not based on how two brains responded to a particular comedian or documentary but instead on the patterns of brain connectivity extracted from periods of task-free mind wandering. And these brain indices even predicted how likely two people were to be socially connected above and beyond similarities in personality measures!

Let's talk about how this brain mirroring might play out in the context of Kanter's model of successful relationships. If our *default* is to understand another mind through mirroring, then pairs of people with more similar brains should get it right more often. This, in and of itself, should increase the number of positive interpersonal interactions they would have, which would reinforce their relationship. If you add the bonus convenience that comes with finding the same environmental stimuli rewarding, it becomes easier and easier to see why *brains* of a feather might flock together.

But what happens when the way you feel in someone else's shoes is not the same way they feel when they're wearing those shoes? How can we explain all of the notable collaborations in history that occur when people who don't see the world through the same lens come together to create a whole that is greater than the sum of its parts? In the next section, we'll discuss another critical mechanism we have that allows us to form a mental model of minds that work differently from our own. As you'll learn, such an ability is critical to being able to connect with others, especially when their brains work differently from your own.

Developing theories of mind

Although they are not our instinctual way of understanding others, most humans eventually acquire more sophisticated ways of reverse-engineering the mind of another than mirroring. But we're not born with these abilities. We have to learn them. If you've ever seen a toddler "hide" by closing their own eyes, you've seen what reasoning about others looks like before we're able to override the mirroring process. In fact, very young children don't even seem *aware* of the fact that different minds have different contents. But eventually, sometime between the ages of two and five,[9] most of us learn that if we close our eyes, *others* can still see *us*, even if *we* can't see anything.

Of course, the extent to which this translates to understanding the more sophisticated points of view of others varies. Because there are many different *aspects* of the minds of others that we might

---

9. The age range given here reflects the fact that some of the tests used to measure perspective-taking are more difficult than others, and hence yield different estimates of what children are able to understand about others and when.

attempt to model. And as you might imagine, being able to understand what the person sitting across from you sees through their own eyes requires a different type of mental process from forming a model of what they might be thinking about. And an increasing amount of evidence suggests that understanding how someone *feels* might be completely independent from either of those perspective-taking exercises. Unfortunately, the same term—*theory of mind*—has been used to explain a variety of different, but related, processes that can be used to model the mind of another person with a different perspective from your own.

To unpack these layers a bit, let's begin by talking about the type of mind modeling that has been studied most frequently through an individual-difference lens—the ability to make inferences about what someone else thinks or knows. One of the most common ways this is studied during development is with a paradigm called the False Belief task.[10] With young children, a typical False Belief experiment goes something like this: The experimenter shows a child some kind of familiar container, like a box of crayons, and asks them what they think is inside. "Crayons!" the child responds enthusiastically. Then the experimenter opens the box and reveals a plot twist—the crayon box is full of birthday candles! Even two- and three-year-old children are shocked by this, evidence of statistical prediction mechanisms hard at work. But here's where things get interesting: The experimenter puts the candles back into the box of crayons and closes the lid. Then they ask the child what someone else who isn't in the room (like a parent or sibling) will think is in the crayon box. Children under the age of four almost always respond, "Candles!"

Of course, there are individual differences in how soon, and how accurately, children learn to represent their own knowledge

---

10. These are fun to watch. If you're interested, you can search YouTube for "False Belief Test: Theory of Mind" to see it in action.

separately from that of others. And since the timing of this ability often coincides with the development of other frontal lobe "control" functions, some researchers have argued that taking the perspective of another person requires *inhibiting,* or overriding, your own perspective. In other words, you can't walk a mile in someone else's shoes without taking your own shoes off first!

To test this hypothesis, Stephanie Carlson and Louis Moses conducted an experiment in which they measured both inhibitory control and mind modeling in more than 100 three- and four-year-old children. Carlson and Moses used a bunch of tests to measure inhibitory control, ranging from resisting real-world temptations, like asking the toddlers "not to peek" while a gift was wrapped behind their backs, to more cognitive tasks, like having them point to a green square when the experimenter said "snow" and a white one when they said "grass." The participants in this study were also given a series of False Belief tests, like the crayon/candle scenario. The results of these tests showed that children who were better able to inhibit automatic responses were also better at modeling the thoughts of others. This evidence is *consistent* with the "taking your own shoes off first" idea; however, because the data are correlational and collected at one time point, there are other possible explanations that can't be ruled out. For instance, being able to model the thoughts of others might have helped children understand what they were supposed to do in the inhibitory control tasks. I imagine that if you're a three- or four-year-old, it would seem pretty strange for an adult to ask you to do something blatantly wrong, like say "day" when they show you a picture of night. It's possible that having some kind of understanding that people can know different things than you do, and might play "tricks," like putting candles in a crayon box, could also help you understand how to play a *game* that involves doing something that's the opposite of what you're used to.

Fortunately, research from the field of behavioral genetics has provided some interesting and complementary evidence about how nature and nurture contribute to the ability to model the thoughts of others. For example, one study conducted by Claire Hughes and colleagues measured mind-modeling abilities in *over 1,000 pairs* of five-year-old twins.[11] The rich data set collected from this large sample painted a crystal-clear picture of the relative roles that nature and nurture play in shaping children's ability to reverse-engineer the thoughts of others. Specifically, when the researchers compared the similarities in mind-modeling abilities of monozygotic (identical) twins, who made up a little more than half of the sample, to those of same-sex dizygotic (fraternal) twins, they found *exactly* the same correlation ($r = .53$) between twins in each group! This provides convincing evidence that the observed similarities between twins are related to their shared environments and not to their shared genetics.

This stands in stark contrast to research that measures individual differences in inhibitory control and other frontal lobe control processes. Because, as the title of a paper by Naomi Friedman and collaborators puts it, "Individual Differences in Executive Functions Are Almost Entirely Genetic in Origin." In their analysis of control tasks collected from 582 pairs of twins, they estimated that a whopping 99 percent of variability in inhibitory control was explained by genetics, leaving only 1 percent to be explained by environmental factors. Taken together, this body of research suggests that though inhibitory control and belief-modeling abilities are related to one another in young children, the mechanisms that shape them are quite different. Given how important understanding one another is, the million-dollar question becomes: What features of an environment might promote the ability to learn to model the mind of others?

---

11. In the spirit of mind modeling, I'd like to note *how impressive* this research effort was. The team tested over 2,200 five-year-olds, in over 3,000 hours of home visits! If you've ever tried to get a five-year-old to comply with your plans, you may have some appreciation of how challenging this must have been.

Learning to speak the language of the mind

It's remarkable to think that something as difficult, and as important, as understanding the contents of someone else's mind might be shaped entirely by our environments. And one aspect of a child's environment that has been consistently linked to their ability to model the thoughts of others is its language content. For instance, Hughes and colleagues also measured verbal abilities in their large study of five-year-old twins. Their sophisticated analysis of these data allowed them to discover that a common environmental factor explained significant variability in both verbal and mind-modeling abilities. This makes sense to me, if you consider the fact that language is one of the "observable" behaviors with the greatest potential to provide clues about what's going on in someone else's mind. For example, as the parent of an extremely empathetic toddler, I could say things to Jasmine like "When you do this, it makes me feel worried because I'm afraid you'll get hurt"—or "I'm stressed out because I'm trying to finish my homework." Because her empathy and mirror neurons knew what it was like to feel worried or stressed out, this helped to guide her behavior. What I didn't realize at the time was that I was also giving her a verbal window into my inner world.

At the end of the day, skilled language users have a tool that allows them to effectively exchange information about what's going on in their brain. But understanding the contents of someone else's mind might also make *you* a more effective language user, because successful communication also requires some understanding of where the brain receiving the signals is coming from. So which comes first, the ability to model the thoughts of another, or the ability to use language effectively?

According to a longitudinal study conducted by Janet Astington and Jennifer Jenkins, the answer is probably language. In their study, they followed a group of three-year-olds over a seven-month period, assessing both language and mind-modeling abilities at three separate points in time. Their results showed that earlier language abilities could be used to predict performance on False Belief tests at later time points, while performance on False Belief tests did not predict subsequent language performance.

Further evidence from the parenting literature provides a more specific link between language environments and how one might learn to model the thoughts of others. Specifically, a series of studies by Elizabeth Meins and her collaborators developed the term *mind-mindedness* as a construct measuring how aware and responsive caregivers are to the minds of their young ones. Meins first described the recursive idea of mind-mindedness in a 2001 study that investigated the predictors of mother and infant attachment. In it, she found that moms who talked about the mental states of their six-month-old infants had more secure attachments with them when they were tested in the lab six months later. Then, in a follow-up study, the critical link was made. Meins showed that the same six-month-old infants who had the more "mind-minded" mothers performed better on False Belief tests three and a half years later. This is a particularly remarkable finding, given what the behavioral genetics research suggests about the lack of a genetic component in mind modeling.

The combined results from longitudinal research and behavioral genetics studies strongly suggest that children who are bathed in rich linguistic environments, and particularly those who have a lot of content about both their own and others' mental states, learn to understand the minds of others sooner. This forms a concrete connection between learning to *think* about mental states and learning to *talk* about them. But every parent and educator knows that some attempts to teach and model behaviors go more smoothly than others.

What might surprise you is the role that inter-brain alignment plays in successful teaching exchanges.

One of my favorite demonstrations of this essentially measured how likely infants were to take the "Yelp-style" recommendations of their parents to heart. In it, Victoria Leong and her collaborators recorded the electrical activity from the brains of 47 pairs of mothers and their ten- to eleven-month-old infants as they engaged in an exchange of information about new objects. The experiment begins with Mom being handed two objects that the infant has never seen before.[12] She then picks up one object and *either* gives it her enthusiastic endorsement—"This is awesome! We like this!"—or does the opposite: "This is yucky! We don't like this!" In the experimental photos included in the article, you can also see that moms' expressions provide consistent information about whether the object is good or bad. The experimenter then passes both objects to the child. To see whether the baby learned anything from their mom's Yelp review, they measured how much time the baby spent playing with the toy Mom commented positively or negatively on versus the one she didn't talk about. As it turned out, stronger synchronization between the brains of Mom and baby during Mom's commentary increased the likelihood that the baby would learn from it. And follow-up analyses showed that both eye contact and length of maternal utterances increased the degree of brain synchronization between pairs during successful interactions. In other words, the more information a baby gets from Mom's face and voice, the more their two brains become synchronized.[13] When this happens, the information exchange between them goes more smoothly.

---

12. They don't talk about these objects much in the paper, but from the pictures, they look like very boring pieces of plastic—the kinds of things infants wouldn't have strong opinions about going into the experiment, which makes sense, given its goal.

13. This likely happens with dads and other caregivers who interact regularly with babies as well, but most studies have focused on moms. This is another space in

But before you go incorporating these practices into your interactions with infants, here's an important "not-so-fun fact" to consider. These ten- and eleven-month-olds didn't always learn by *mimicking* their moms' recommendations. In fact, the more different the temperaments of mother and baby were, as measured by parent reports, the more likely the baby was to pick up the object their mom did not recommend! The fact that they did this consistently suggests that they *were learning* from their parents' reactions, but that somehow their tiny brains factored in the similarity between themselves and their mothers when deciding *what to do* with that information.[14] The fact that ten-month-olds may already be making decisions about whether to follow their mom's advice nicely illustrates the point we'll discuss in the next section: Not everyone is equally *motivated* to do the work it takes to become aligned with someone else's point of view!

Motivating connection

So far, most of the research we've talked about has focused on the conditions that promote *learning* the skills necessary to understand the thoughts of others. But as we learned in "Navigate," just because someone *knows how* to do something, it doesn't necessarily mean that they *will* do it. Given how difficult it is to understand invisible, unpredictable human minds, and how *bad* the consequences of misunderstandings can be, what motivates us to try?

---

which our systemic biases become represented in our science, because—at least in the United States—moms are far more likely to be the caregivers who are willing and able to bring their children into the lab for research. Shout-out to my lab manager, Justin, who is taking six months of paternity leave to take care of his kids while his wife returns to work!

14. While I bet that there are some serious "Aha!" moments happening in the brains of the parents reading this book, I'm also hoping that those of you who don't have children are gaining insights about your interactions with your parents.

The answer to this question will take us all the way back to some of our earliest discussions of the neuroscience of you based on your brain's smallest design features. Lying there, among the ingredients of your neural cocktail, is *oxytocin*—the neurotransmitter most strongly implicated in promoting social attachment in mammals. To get us to put in the *work* required to have successful relationships, oxytocin can both stimulate dopamine-receiving neurons to enhance the pleasure experienced during social interactions and also influence the amygdala and other fight-or-flight limbic regions in ways that *reduce* the brain's natural stress response to social interactions. In other words, when oxytocin is on board, it can decrease the likelihood that your brain's appraisal system will experience threat when approaching another person, which increases the chance that you will explore a connection with them.

Much of what we know about oxytocin's role in promoting social connections stems from measuring changes *within* a person or animal during significant relationship milestones. Becoming a parent is among the most salient of these milestones—one that involves *a lot* of hard work, and also happens to be critical for survival of the species. But if you think understanding the mind of another full-grown person is hard, try being "mind-minded" about a baby, especially during the "blooming, buzzing confusion" phase. From an evolutionary standpoint, this would be a great time to provide a little *nudge* to help someone care about another person's perspective, wouldn't it?

In fact, this is one of the critical roles that oxytocin plays in keeping mammals on the planet. In the female body, oxytocin functions as a hormone that induces labor, and is released during lactation. In the brain, oxytocin levels increase in both mothers and fathers when they touch and engage in social interactions with their infants, increasing feelings of connectedness and decreasing stress. Longitudinal data suggests that oxytocin levels continue to rise in parents for at least the first six months of an infant's life, and that parents who cohabit have correlated oxytocin levels.

But parents aren't the only ones in these relationships that need a strong dose of oxytocin. As you might imagine, being born helpless into a world full of things you can't make sense of is pretty freaking stressful for a baby. Most of the other *animals* around them are giants, and their "Spidey senses" must detect that these giants are capable of harming them. Yet the giants also seem able to provide nourishment and to improve their comfort levels more generally. The only tools babies have to deal with this drama are a few reflexes and a brain that can learn a lot in a really short time. To enhance their chances of survival, the infant brain needs to learn, quickly, which of the giants they should *trust*. Because, even though the infant won't be capable of running away for another year or so, they are able to smile, and coo, and engage in a host of cute and progressively more complicated behaviors that can motivate the giants of their choice to take care of them.

Research on both human and nonhuman animals suggests that oxytocin plays a critical role in this early bonding process for infants. Remarkably, newborn lambs and their mothers provide a very interesting model of the role of oxytocin in forming connections. Unlike humans, lambs are born into a herd and are on the move shortly after birth. These environments create a forcing function for figuring out which of the big woolly creatures is "Mama" ASAP! And within their first two hours of life, lambs that have the opportunity to suckle will begin to both recognize and prefer their biological mothers to the other giants around them. A recent study, published in 2021 by Raymond Nowak and colleagues, provided converging evidence of oxytocin's role in this early lamb bonding process. In a series of experiments with newborn lambs, the research team first showed that oxytocin levels in lambs rose after suckling, but not after other, non nutritive interactions with Mom. Then they showed that newborns that were given a drug that *blocks* oxytocin binding in the brain both explored their mothers' bodies less frequently and showed less of a preference for her in the short term.[15]

Though human newborns don't have the opportunity to walk away and get lost in a herd, the limited research we do have about their oxytocin levels suggests that they play a similar role in parental bonding. For instance, one study conducted on preterm newborns showed that skin-to-skin contact with either parent increased oxytocin levels in both parent and child. And to link back to the *health* benefits of such intimacy, skin-to-skin contact was also shown to decrease the infants' cortisol[16] levels. Taken together, this sampling of the research on parent-child bonding suggests that—*on average*—changes in oxytocin levels within an individual coincide with critical bonding moments. But before children can come into existence—at least in the traditional way—their parents also have to connect!

As it turns out, oxytocin is also involved in motivating sexual and romantic relationships—and we've got voles[17] to thank for most of what we know about it. In 1992, Thomas Insel and Lawrence Shapiro went looking for the biological basis of monogamy in the brains of two species of voles that resemble each other in many ways, with the exception of their social practices: prairie voles and montane voles. In the wild, prairie voles tend to form long-term, monogamous relationships, and both sexes are involved in caring for their young. Montane voles, on the other hand, live in isolation, are not known to be monogamous, and spend very little time caring for their pups.

---

15. This is *sad*, but don't worry, the effects of the drug on both the brain and the bonding were temporary. All changes disappeared within forty-eight hours, reuniting the lambs to the woolly giants that birthed them.

16. As you probably remember, cortisol is a neurotransmitter that's related to prolonged stress experiences.

17. Voles are adorable rodents. They look a little bit like a punk-rock hamster.

When Insel and Shapiro investigated the binding patterns of three different neurotransmitters in the brains of the two species of voles, they found whopping differences in their oxytocin communication systems. The monogamous prairie voles had more oxytocin receptors in six out of the ten brain regions investigated, including more than *six times* as many receptors in the nucleus accumbens, the part of the basal ganglia most strongly associated with receiving dopamine projections and feeling pleasure. They also had more than twice as many oxytocin receptors in the lateral amygdala, which is involved in the fight-or-flight response. A series of subsequent follow-up studies extended this work with causal manipulations. Oxytocin delivered to prairie voles before periods of nonsexual cohabitation increased their preference for each other, while a drug that blocked oxytocin binding did not interfere with mating but did prevent the prairie voles from forming partner preferences afterward. In other words, at least in monogamous mammals, oxytocin seems to be involved in driving adults to connect, much like it does in parent-infant relationships.

These findings quickly spawned a bunch of interesting research on the role of oxytocin in adult human relationships. For example, in a clever series of experiments, Dirk Scheele and colleagues gave oxytocin to human participants and measured its effects on their brain and behavior. In two similar studies, Scheele and collaborators examined the effects of oxytocin on the brain responses of 40 men in committed relationships who, according to their self-report measures, were "passionately in love." In the scanner, each man viewed pictures of their partner, pictures of familiar women who were not related to themselves or their partner, pictures of strangers (selected by independent raters to match those of their partners on attractiveness and arousal levels), and pictures of neutral stimuli, like houses. To measure the effect of oxytocin on their perception of their partners, all men underwent two scanning sessions, one with and one without oxytocin. The results, in a nutshell, showed that oxytocin made men's brains more like the brains of monogamous prairie voles. To be more specific,[18] both experiments showed that when oxytocin was on board, brain activation increased in the nucleus accumbens,[19] the dopamine-sensitive reward center where prairie voles have massive amounts of oxytocin receptors. What was really remarkable about these findings is that this *only* happened when men saw pictures of their partners—*not* when they saw pictures of other women, strange or familiar, that were objectively equally attractive.[20]

In a clever follow-up study, Scheele and collaborators explored the "real world" effects of oxytocin administration using a paradigm called the "stop distance" task. The task involved several conditions in which a study participant was positioned face-to-face with an experimenter and was asked to decide the distance at which they felt comfortable standing. Sometimes the experimenter would start far away and move toward the participant, and other times they would start close and back away. In both cases, the participant would tell them to "stop" when they reached a comfortable distance. In other conditions, the participants would approach the experimenter from far away, or start close and back up. In both of these cases, the participants simply stopped themselves when they were at a comfortable distance. Then, at the end of each trial, the final distance between the pair was recorded from chin to chin.

And here's where things get interesting. All of the participants in this experiment were heterosexual men, and the experimenter was an attractive woman, as rated by an *independent* group of people. And

---

18. Especially since prairie voles have *tiny* brains . . .

19. Both experiments also showed increased activation in the ventral tegmental area, the region of the basal ganglia that releases dopamine in the face of rewards.

20. I don't want to kill the mood, because this is very sweet, but it *is* worth pointing out that these data were averaged across the men in the group. Can you imagine how juicy the reality-TV version of *The Brain Wants What It Wants* would be? Pump your partner full of oxytocin and show them pictures of your face and the face of equally attractive strangers and see how selective their brains are!

about half of the 57 men who completed the experiment were in stable monogamous relationships and the other half were single. So, what do you think happened when oxytocin was administered to the two groups?

If you guessed that oxytocin would make men in relationships stand farther away from attractive women, you nailed it! With oxytocin on board, men in committed relationships stood about fifteen centimeters farther away from the attractive experimenter. Without oxytocin, however, men in relationships felt comfortable at single-man levels of closeness. Taken together, these results suggest that experimentally adding oxytocin to the mix increases the perceived reward value of a partner, which can motivate men to behave in more selective, pair-bonded ways.

Recently, Simone Shamay-Tsoory and Ahmad Abu-Akel helped to provide a mechanism for how oxytocin influences social bonding by proposing that its actual computation is to *enhance* the salience of socially relevant information in the environment. In "Focus," we talked about how the basal ganglia do this—effectively turning the signals traveling to the prefrontal cortex up or down based on what they think is relevant. According to the *social salience hypothesis*, oxytocin receptors in the basal ganglia can essentially *hijack* this process and turn up the volume on socially relevant signals.

If you think about it, this is an evolutionarily smart thing to do. Rather than having a human baby focus on the first thing they lay eyes on, they are given the tools to help them learn to understand the things that they have the most potential to benefit from. This would allow important signals from caregivers to get turned up against the blooming, buzzing confusion of the rest of the world so that the babies can learn whom to trust.[21] This might also explain why

---

21. In fact, human babies do have built-in preferences for attending to socially relevant stimuli like faces and voices.

ten-month-olds already have enough data about their caregivers to decide whether they agree with their opinion about a shiny new object.

Consistent with this idea, a series of experiments suggest that increased oxytocin levels are correlated with improved mind-modeling abilities. However, the results of studies investigating this relationship have been inconsistent. For example, one experiment by Gregor Domes and colleagues found that males who were administered oxytocin performed better on the most difficult items in the Eyes Test. Another study by Sina Radke and colleagues, which followed a very similar protocol, did not find any improvements at the group level, but did find that men who scored lowest on a self-rating scale for the trait of empathy *did* improve on the Eyes Test when given oxytocin. Finally, a meta-analysis that analyzed results across multiple studies of emotion-reading with oxytocin suggested that oxytocin might *only* help with the recognition of a handful of amygdala-related emotions, like fear or anger. One possible explanation for these results, discussed in a commentary written by Jennifer Bartz and collaborators, shouldn't surprise you one bit. They describe the likelihood that oxytocin *doesn't do the same thing in everyone,* and this is because *different* information is socially relevant in specific contexts and for specific individuals.

This makes a lot of sense, given everything else you've learned in this book. The way you focus, combined with your lived experiences, is bound to shape the kinds of clues your brain learns to use when trying to navigate social situations. And here's where things get disappointing—at least from my perspective. It doesn't seem like oxytocin is the magic ingredient that will bring people who are *different* from one another close together. In fact, it probably does the opposite.

In yet another twist on the costs and benefits of different mechanisms of connecting, research suggests that the very chemical that motivates us to connect seems to enhance our awareness of "in-group" and "out-group" differences. For instance, a series of studies by Carsten de Dreu and collaborators showed that administering oxytocin to adult males increased their ethnocentric, or in-group, biases. Another study showed that giving oxytocin to adult males[22] increased their sensitivity to painful expressions, but only on same-race faces.

One thing to keep in mind is that oxytocin doesn't *create* these in-group biases. It is much more likely to *enhance* the salience of biases that already exist. In both experiments, men in the placebo groups already showed in-group biases. The fact that their biases were turned up following oxytocin probably reflects an "amplification" of some preexisting social cue that their brains already use to divide the world into "us" and "them."

Consistent with this idea, when Michaela Pfundmair and her research team assigned 60 male and female participants to groups that were *allegedly* based on their art preferences, oxytocin did *not* enhance in-group biases. In the study, the researchers first presented their participants with several pairs of paintings and asked them to choose which one they preferred. After ten pairs of paintings were presented, each participant was told that they preferred paintings by "Pechstein," no matter what their actual choices were. They were then assigned to "Team Pechstein." Unbeknownst to them, there was no other team, but they were led to believe in a fictional "Team Heckel," which consisted of people who liked different paintings than they did. After being assigned to teams, participants watched a series of very boring-looking videos (the paradigm was created for

22. If you're wondering why so many of these studies were conducted only on men, you're not alone. My best guess is that there are reasons based on biology, social roles, or both, that men and women might differ in these responses. And if you don't have enough money, or stimuli, to run both, you run men because— nope, I can't scientifically justify anything further than that.

infants) in which either a hand or mechanical grabber arm reached onto a screen and moved toward one of two objects. Before each video with a human arm, a note on the screen said whether they belonged to Team Pechstein or Team Heckel.

Note, although it's incredibly unlikely that participants walked into this study with any kind of meaningful identity associated with either of the painters, being assigned to Team Pechstein caused people to rate themselves as feeling more empathetic toward other members of Team Pechstein than they did toward members of Team Heckel. But the addition of oxytocin *didn't* enhance this effect. Instead, people given oxytocin spent more time looking at videos with human hands from either team than they did looking at the mechanical grabber arms. And though people both on and off oxytocin looked slightly longer at hands from Team Pechstein, the effect was neither significant nor significantly bigger following oxytocin. In short, if life hadn't already taught them that there was a benefit for affiliating with people who liked Pechstein, oxytocin didn't create one.

Perhaps my optimistic, dopamine-driven brain is playing tricks on me, but I think this leaves room for hope when it comes to our ability to connect with others across our differences. While the results of these studies suggest that oxytocin may enhance the social cues that are already associated with in-group/out-group distinctions, they also leave space for learning which are the important cues to *you*, should you decide to redefine what counts as your pack.[23] And in case the importance of this isn't as obvious to you and your brain as it is to me and mine, in the next section, I'm going to cover some

---

23. The metaphor is particularly appropriate, since oxytocin has consistently been shown to increase reciprocally when humans look at their dogs and when dogs look at their humans. Isn't it beautiful how widely we *can* cast our nets of belonging?

of the measurable benefits that come with being able to reverse-engineer the mind of another person.

Putting the eyes in team

So far in this chapter, we've discussed how our innate mechanisms for understanding the actions of others are egocentric, and how this may drive people to spend time with other like-brained individuals. But we also read some pretty strong evidence suggesting that both the ability to understand the minds of others and the social cues we consider when doing so are *learned*. So before I let you close the book on the idea of connecting with people who work differently from you, allow me to make a case for the idea of a *collective social intelligence*, and the role that mind-modeling has been shown to play in it. Just in case, to paraphrase Maya Angelou, knowing better helps to lay the groundwork for doing better.

Whether we like it or not,[24] there are times in life when we are asked to cooperate with groups of people we don't get to choose. From classroom projects to the workplace, team membership is an essential part of the human experience. And when things go *well* in teams, the effect is that the performance of the team is truly greater than the sum of its parts. So it's no surprise that organizational psychologists have spent decades trying to understand the "magic recipe" for forming a successful team.

In the past decade, the science of teamwork has taken a leap forward, due—in no small part—to the consideration of mind-modeling abilities. Two enormous teamwork experiments led by Anita Woolley and colleagues provide clear examples of this progress. The first thing Woolley and her team did was to figure out a way to measure

---

24. Which probably has a lot to do with how extraverted you are . . .

the success of a team using a term she defined as their "collective intelligence." To do so, she first randomly assigned 699 people to groups ranging from 2 to 5 people. Then she asked the groups to collectively work to solve a wide variety of problems selected to assess team performance under a number of different conditions. The problems ranged from solving visual puzzles to making moral judgments or negotiating how to use limited resources. Then these teams made up of total strangers worked together to achieve these different, experimenter-defined goals for up to five hours.

One of Woolley's first critical contributions to the field was to show that successful teams performed better than less-successful teams *independently* of the task they were asked to do. In other words, it's not that the randomly assembled groups of people sorted themselves into "puzzle teams" or "logistical teams." Instead, the *collective intelligence* measure Woolley derived for each team, based on their performance on *all* of the tasks, explained over 40 percent of their success across all problem-solving outcomes when compared to other teams. Based on this measure, she was then able to ask the question researchers in the field had been studying for years.

What predicts the success of a given team?

One of the biggest surprises was that *neither* the average intelligence of the individuals on the team nor the maximum intelligence of any individual on a team was strongly related to collective intelligence.[25] Neither were measures of average team motivation, satisfaction, or group cohesion. Instead, three factors reliably predicted how well teams of people would perform across the board: (1) their average abilities to reverse-engineer the human mind, as measured by the Eyes Test, with better performance on the test being associated with

---

25. In each of the two individual experiments reported, these factors were not significantly correlated with collective intelligence. However, when the results from two experiments were averaged, they did reach significance, explaining 2.2 percent and 3.6 percent of the variance, respectively.

better performance by the team; (2) the distribution of spoken turn-taking, with more distributed turn-taking being associated with better team performance; and (3) the proportion of female participants on any team, with a larger number of females being associated with better team performance.[26] These revolutionary findings were backed up by further research showing that performance on the Eyes Test also predicts team performance on classroom projects, as well as in *online collaborative settings*. In the final section of this chapter, I'll do my best to use language to share a bit of the contents of my mind with you, as we connect the dots between what we've learned about how we understand others.

Summary: Inter-brain similarity drives success of mind mirroring, while experience with "mind-mindedness" shapes our ability to model others

When you tie the body of results discussed in this chapter together, the implications are pretty profound. First, the most instinctual way we have of understanding others is to put ourselves in their shoes. And though this can create a powerful, empathetic, "feeling with" experience, it doesn't necessarily work well when the person you're trying to connect with wears a different size shoe than you do. And since not everyone works the same, using these "mirroring" mechanisms probably drives the homophily we see around us. People whose brains work in similar ways tend to assemble into like-minded packs.

But there are other, more effortful ways to reverse-engineer another's brain based on the observable cues. These "theory of mind" methods seem to be heavily related to a person's ability to use language. Remarkably, they also seem to be entirely learned. But

---

26. I don't make the news, people, I just report it.

knowing what someone else might be thinking or feeling doesn't necessarily mean you'll use that information to behave in a way that takes their feelings into account. The neurotransmitter oxytocin may provide one of the motivational signals for doing so, by enhancing the dopamine reward systems that drive you to want to connect and hence turn up the volume on the socially relevant signals—at least when you interact with groups of people you feel motivated to connect with.

Finally, the teamwork literature suggests that people who are good at reverse-engineering the minds of others are *more successful* in a variety of collaborative environments. And since performance on the Eyes Test also predicts successful online collaborations where teammates couldn't *see* one another, it's likely that the ability to reverse-engineer what someone else is feeling based on their facial expressions reflects a more general expertise in using observable social clues, including language, to *infer* the contents of another's mind. Because these abilities appear to be learned, there should be an opportunity to get better at them. This is good news, because according to Kanter's model, interpersonal relationships depend on the ability to communicate and be understood both verbally and nonverbally. The moral of this story, at least from my brain's perspective, is that practice makes perfect when it comes to mind modeling, and that practice can have tangible benefits in the success of your interactions with others.

And speaking of our interactions with others, I sincerely hope you've learned as much about *yourself* by reading this book as I did when I wrote it for you. If so, as you move forward with all of these new ideas in your saddlebag, I'd like to challenge you to do something *different* when you try to understand yourself and others. It's a task I really hope that reading this book will leave you better prepared to navigate.

Can you reverse-engineer your thoughts, feelings, and behaviors based on what you now know about how brains work? Even if the idea might sometimes make you feel uncomfortable, does this new understanding of how your brain builds your reality change the way you understand yourself?

And now, can you take this exercise one step further and try to understand why someone else, who could be behaving in a way you find totally idiotic, might just be driven by a different brain that has been shaped by different experiences? Since the brain is the creator of the mind, I firmly believe that doing so will give you the most powerful mind-modeling abilities possible. After all, trying to walk a mile in another person's shoes is bound to give you blisters if they aren't the right size. Instead, I hope you'll join me on this adventure and try walking a mile in their brain. Because a peek inside our different biological ways of being might open *your* mind to a world, as of yet, unexplored.

ACKNOWLEDGMENTS

When I was young, I used to joke that I would write a book someday and dedicate it to Norah—the Girl Scout leader in "liberal on the outside but conservative on the inside" Davis, California, who made Jasmine and me feel like crap because we didn't fit in with the other soccer moms in her troop. But twenty years have passed, and I'm slightly less pissed about it now. In fact, I am *almost* grateful for the way that experience fueled my growth, which eventually motivated me to try to understand *why* someone would behave that way. So even if my motivation for doing so is *slightly* less childish than I originally envisioned—thank you, Norah.

But I'd prefer to focus on those who have supported me, who thankfully outnumber the naysayers a thousand to one. I'm *so* lucky. So I'll start my *sincere* thanks with the lifelong supporters—my parents. My interest in individual differences must stem from the fact that my mom and dad are two of the most different people I've ever met. I can't imagine anything short of a small town and the Free Love movement could've brought them together, but I'm glad it did, even if only for a little while. As you might imagine—I was not the easiest child to parent. Thanks, Momma, for always doing your best to parent me even though we don't quite "work" the same. Thanks to my dad for encouraging me to dream big. And thanks to my stepdad, Jim, for giving me all the car skills to make these metaphors, and for vetting several versions of them before they appeared in the book.

And then, of course, there's the best friend I gave birth to— Jasmine. I remember the moment I dreamed of what you could be, and how surprised I continue to be about how much *more* you are than that. Thank you for caring about me, and for teaching me how powerful connection can be. And thank you for your slow, but thorough, insightful, and *hysterical* comments on the first third of the book. I hope you like the rest when you get a chance to read it. ☺

Which brings me to the first person who read this book in its entirety—Andrea. From your illustrations to every one of the walks and hikes we took where we exchanged book ideas to the fact that you almost entirely took over household responsibilities while I was buried in two full-time jobs—there is no way I could have done this without you. I'm a little worried about telling a bunch of perfect strangers how great you are, but I'm counting on oxytocin, plus the fact that no one can basal ganglia like I can, to keep us connected, even if you're way too good for me. Thank you for being my biggest supporter. I wish I could see the world (and myself) through your brain.

Next, to my agent extraordinaire, Margo Beth Fleming, and my brilliant editor, Jill Schwartzman—talk about a power couple. Thank you for taking a chance on someone with very little to show for herself but ideas . . . Man, did I take you guys on a bumpy ride. And thanks for explaining *everything* to me, for listening to me, and for your appreciation of the ridiculous memes and photo updates I sent with each check-in.

I would also like to thank Ray Perez and the Cognitive Science of Learning Program at the Office of Naval Research, who currently fund my research and supported some of my book writing. And thanks to the Sou'wester Lodge's Artist Residency Program for housing me and my dog during some of the heaviest intellectual lifting involved in writing this book.

And of course, thanks to all the friends, family, and students who either read parts of the book or gave me ideas about what would be interesting to talk about: Eddie, Jen K., Jenni, Jeanne, Brianna, Katie (my muse), David & Judy (the family I found in the process of writing), Caitlin, Cousin Danny, Shaya, Uncle Danny, Jen J., Kira, Maria, Michelle, Aunt Jan, Tonya, Richard, Stacie, Robin, Holy A., Julie, Uncle Larry, Kristy, Annamarie S., Claire, Deanna, Deborah, Jeffrey, Obadiah, Kim (RIP), Dawn, Dina, Charlotte, Erik, Rabiah, Akira, Yinan, Olga, Zirui, Malayka, Lauren, Thea, Marissa, Margarita, Jim, Jay, Cher, Preston, Amanda, Mari, and Shreya. Extra-special thanks to Justin and Jim for their "eagle-eyed" proofreading. It takes a village, and my village is strong.

And last, but definitely not least, to my dog, Coccolina, who sat by my side for 95 percent of this writing. Thank you for flooding my brain with oxytocin and for teaching me that sometimes just being next to someone is the best thing you can do for them.

NOTES

PREFACE

xi   **fundamentally, and permanently, changed:** Kieran O'Driscoll and John Paul Leach, "'No Longer Gage': An Iron Bar Through the Head: Early Observations of Personality Change After Injury to the Prefrontal Cortex," *British Medical Journal* (1998): 1673–1674.

xi   **"fitful, irreverent":** John M. Harlow, "Recovery from the Passage of an Iron Bar Through the Head," *History of Psychiatry* 4, no. 14 (1993): 274–281.

xi   **a neurochemical related to prolonged stress:** David M. Lyons et al., "Stress-Level Cortisol Treatment Impairs Inhibitory Control of Behavior in Monkeys," *Journal of Neuroscience* 20, no. 20 (2000): 7816–7821.

INTRODUCTIONS

3   **2 percent of its total weight:** See, for example, Marcus E. Raichle and Debra A. Gusnard, "Appraising the Brain's Energy Budget," *Proceedings of the National Academy of Sciences* 99, no. 16 (2002): 10237–10239.

3   **folded in on themselves:** Este Armstrong et al., "The Ontogeny of Human Gyrification," *Cerebral Cortex* 5, no. 1 (1995): 56–63.

3   **same surface area as two medium pizzas:** David C. Van Essen et al., "Development and Evolution of Cerebral and Cerebellar Cortex," *Brain, Behavior and Evolution* 91 (2018): 158–169.

4   **"Big-Brained People Are Smarter":** Michael A. McDaniel, "Big-Brained People Are Smarter: A Meta-Analysis of the Relationship Between In Vivo Brain Volume and Intelligence," *Intelligence* 33, no. 4 (2005): 337–346.

4   **"intelligence is what intelligence tests test":** Edwin G. Boring, "Intelligence as the Tests Test It," *New Republic* 35, no. 6 (1923): 35–37.

7   **The brains of London taxi drivers:** Eleanor A. Maguire et al., "Navigation-Related Structural Change in the Hippocampi of Taxi Drivers," *Proceedings of the National Academy of Sciences* 97, no. 8 (2000): 4398–4403.

7   **fewer than 50 percent:** Katherine Woollett and Eleanor A. Maguire, "Acquiring 'the Knowledge' of London's Layout Drives Structural Brain Changes," *Current Biology* 21, no. 24 (2011): 2109–2114.

8   **London bus drivers:** Eleanor A. Maguire, Katherine Woollett, and Hugo J. Spiers, "London Taxi Drivers and Bus Drivers: A Structural MRI and Neuropsychological Analysis," *Hippocampus* 16, no. 12 (2006): 1091–1101.

10 *five* **or more symptoms:** American Psychiatric Association, *Diagnostic and Statistical Manual of Mental Disorders (DSM-5®)* (American Psychiatric Publishing, 2013).

11 **You might decide for yourself:** Edward M. Hallowell, MD, and John J. Ratey, *Driven to Distraction: Recognizing and Coping with Attention Deficit Disorder from Childhood Through Adulthood* (New York: Anchor, 2011).

13 **are WEIRD:** Joseph Henrich, Steven J. Heine, and Ara Norenzayan, "The Weirdest People in the World?" *Behavioral and Brain Sciences* 33, no. 2–3 (2010): 61–83, and Joseph Henrich, *The Weirdest People in the World: How the West Became Psychologically Peculiar and Particularly Prosperous* (New York: Farrar, Straus and Giroux, 2020).

15 **"As human beings":** Fred Rogers, *You Are Special: Neighborly Words of Wisdom from Mister Rogers* (New York: Penguin, 1995).

15 **"all *normal* people":** Steven Pinker, *How the Mind Works* (Penguin UK, 2003).

15 **"Differences among people":** Pinker, *How the Mind Works.*

16 **132 muscles and 26 organs:** John G. White et al., "The Structure of the Nervous System of the Nematode Caenorhabditis Elegans," *Philosophical Transactions of the Royal Society of London, Series B, Biological Sciences* 314, no. 1165 (1986): 1–340. Also see Steven J. Cook et al., "Whole-Animal Connectomes of Both Caenorhabditis Elegans Sexes," *Nature* 571, no. 7763 (2019): 63–71.

17 **a quick search of *action potential*:** (There are many of these, but here's one I like!) "Action Potentials in Neurons, Animation," Alila Medical Media, YouTube video, uploaded April 25, 2016, https://www.youtube.com/watch?v=iBDXOt_uHTQ.

18 **dozens of books:** Cornelia I. Bargmann, "Neurobiology of the Caenorhabditis Elegans Genome," *Science* 282, no. 5396 (1998): 2028–2033; Anders Olsen and Matthew S. Gill, eds., *Ageing: Lessons from C. Elegans* (Springer International Publishing, 2017); and Lisa R. Girard et al., "WormBook: The Online Review of Caenorhabditis Elegans Biology," *Nucleic ACIDS RESEARCH* 35, no. suppl_1 (2007): D472—D475.

18 **human and chimp brains overlap by about 95 percent:** Roy J. Britten, "Divergence Between Samples of Chimpanzee and Human DNA Se-

quences Is 5%, Counting Indels," *Proceedings of the National Academy of Sciences* 99, no. 21 (2002): 13633–13635.

20 **it was generally believed:** Debra L. Long and Kathleen Baynes, "Discourse Representation in the Two Cerebral Hemispheres," *Journal of Cognitive Neuroscience* 14, no. 2 (2002): 228–242.

20 **not *all* people understand:** Debra L. Long, Brian J. Oppy, and Mark R. Seely, "Individual Differences in Readers' Sentence- and Text-Level Representations," *Journal of Memory and Language* 36, no. 1 (1997): 129–145.

20 **in more than 200 readers with different skill levels:** Chantel S. Prat, Debra L. Long, and Kathleen Baynes, "The Representation of Discourse in the Two Hemispheres: An Individual Differences Investigation," *Brain and Language* 100, no. 3 (2007): 283–294.

23 **language difficulties following right-hemisphere damage:** John Hughlings Jackson, "A Study of Convulsions," *St. Andrews Medical Graduates' Association Transactions 1869* (1870): 162–204.

25 **Katherine Woollett and Eleanor Maguire did exactly this:** Woollett and Maguire, "Acquiring 'the Knowledge.'"

28 *Three Identical Strangers*: Tim Wardle, dir., *Three Identical Strangers*, Neon, 2018.

29 **about 40 percent of the market:** G. Ferris Wayne and G. N. Connolly, "How Cigarette Design Can Affect Youth Initiation into Smoking: Camel Cigarettes 1983–93," *Tobacco Control* 11 (2002): i32–i39.

29 **likelihood that they will develop dependence on nicotine:** Jacqueline M. Vink, Gonneke Willemsen, and Dorret I. Boomsma, "Heritability of Smoking Initiation and Nicotine Dependence," *Behavior Genetics* 35, no. 4 (2005): 397–406.

31 **4.6 million Americans ride horses:** V. E. Ellie et al., "U.S. Horseback Riders," *Wonder*, 2019, askwonder.com.

36 **adults use language differently:** Jeffrey Z. Rubin, Frank J. Provenzano, and Zella Luria, "The Eye of the Beholder: Parents' Views on Sex of Newborns," *American Journal of Orthopsychiatry* 44, no. 4 (1974): 512. And a follow-up twenty years later: Katherine Hildebrandt Karraker, Dena Ann Vogel, and Margaret Ann Lake, "Parents' Gender-Stereotyped Perceptions of Newborns: The Eye of the Beholder Revisited," *Sex Roles* 33, no. 9 (1995): 687–701.

37 **"People are hard to hate":** Brené Brown, *Braving the Wilderness: The Quest for True Belonging and the Courage to Stand Alone* (Random House, 2017). This is one of my favorite books of all time and it has had a profound influence on me.

PART 1: BRAIN DESIGNS

41 **"A good storyteller weaves":** Brian Levine, "Autobiographical Memory and the Self in Time: Brain Lesion Effects, Functional Neuroanatomy, and Lifespan Development," *Brain and Cognition* 55, no. 1 (2004): 54–68.

CHAPTER 1: LOPSIDED

43 **engineered this way for hundreds of millions of years:** Peter F. Mac-Neilage, Lesley J. Rogers, and Giorgio Vallortigara, "Origins of the Left & Right Brain," *Scientific American* 301, no. 1 (2009): 60–67, http://www.jstor.org/stable/26001465.

43 **which hemisphere is "in charge":** J. A. Nielsen et al., "An Evaluation of the Left-Brain Vs. Right-Brain Hypothesis with Resting State Functional Connectivity Magnetic Resonance Imaging," *PloS one* 8, no. 8 (2013), e71275.

45 **a study that looked at language** *laterality*: S. Knecht et al., "Degree of Language Lateralization Determines Susceptibility to Unilateral Brain Lesions," *Nature Neuroscience* 5, no. 7 (2002): 695–699.

46 *shrinks* **parts of the right hemisphere:** M. Annett, "Handedness and Cerebral Dominance: The Right Shift Theory," *Journal of Neuropsychiatry and Clinical Neurosciences* 10, no. 4 (1998): 459–469; and Marian Annett, *Left, Right, Hand and Brain: The Right Shift Theory* (Psychology Press, UK, 1985).

47 **our long thumbs:** Fotios Alexandros Karakostis et al., "Biomechanics of the Human Thumb and the Evolution of Dexterity," *Current Biology* 31, no. 6 (2021): 1317–1325.

48 **you'd be able to identify:** T. A. Yousry et al., "Localization of the Motor Hand Area to a Knob on the Precentral Gyrus. A New Landmark," *Brain: A Journal of Neurology* 120, no. 1 (1997): 141–157.

48 **by comparing the size of the two knobs:** Katrin Amunts et al., "Asymmetry in the Human Motor Cortex and Handedness," *Neuroimage* 4, no. 3 (1996): 216–222.

48 **Edinburgh Handedness Inventory:** Richard C. Oldfield, "The Assessment and Analysis of Handedness: The Edinburgh Inventory," *Neuropsychologia* 9, no. 1 (1971): 97–113.

51 **2 out of 3 people prefer their right eyes:** D. C. Bourassa, "Handedness and Eye-Dominance: A Meta-Analysis of Their Relationship," *Laterality* 1, no. 1 (1996): 5–34.

53 **how consistently a person depends on one hemisphere:** See, for example, Jerre Levy et al., "Asymmetry of Perception in Free Viewing of Chimeric Faces," *Brain and Cognition* 2, no. 4 (1983): 404–419.

54 **probably picked the face on the bottom:** Victoria J. Bourne, "Examining the Relationship Between Degree of Handedness and Degree of Cerebral Lateralization for Processing Facial Emotion," *Neuropsychology* 22, no. 3 (2008): 350.

54 **your damned finger bounces:** Bourassa, "Handedness and Eye-Dominance."

55 **the decision about which face to pick:** Bourne, "Examining the Relationship." Also see S. Frässle et al., "Handedness Is Related to Neural Mechanisms Underlying Hemispheric Lateralization of Face Processing," *Scientific*

Reports 6 (2016): 27153; Roel M. Willems, Marius V. Peelen, and Peter Hagoort, "Cerebral Lateralization of Face-Selective and Body-Selective Visual Areas Depends on Handedness," *Cerebral Cortex* 20, no. 7 (2009): 1719–1725; and Michael W. L. Chee and David Caplan, "Face Encoding and Psychometric Testing in Healthy Dextrals with Right Hemisphere Language," *Neurology* 59, no. 12 (2002): 1928–1934.

56 **primarily the younger ones or later talkers:** Debra L. Mills, Sharon Coffey-Corina, and Helen J. Neville, "Language Comprehension and Cerebral Specialization from 13 to 20 Months," *Developmental Neuropsychology* 13, no. 3 (1997): 397–445.

57 **occur most often in *extreme* left-handers:** Stefan Knecht et al., "Handedness and Hemispheric Language Dominance in Healthy Humans," *Brain* 123, no. 12 (2000): 2512–2518.

59 **a patient who seemed to lose only the ability to speak:** P. Broca, "Remarks on the Seat of the Faculty of Articulated Language, Following an Observation of Aphemia (Loss of Speech)," *Bulletin de la Société Anatomique* 6 (1861): 330–357.

60 **may be even more important for fluent speech:** Nina F. Dronkers, "A New Brain Region for Coordinating Speech Articulation," *Nature* 384, no. 6605 (1996): 159–161.

60 **to study the preserved brains of Broca's first patients:** Nina F. Dronkers et al., "Paul Broca's Historic Cases: High Resolution MR Imaging of the Brains of Leborgne and Lelong," *Brain* 130, no. 5 (2007): 1432–1441.

60 **lose the ability to use word order:** Myrna F. Schwartz, Eleanor M. Saffran, and Oscar S. Marin, "The Word Order Problem in Agrammatism: I. Comprehension," *Brain and Language* 10, no. 2 (1980): 249–262.

60 **difficulty understanding what actions are depicted:** Ayşe Pınar Saygın et al., "Action Comprehension in Aphasia: Linguistic and Non-Linguistic Deficits and Their Lesion Correlates," *Neuropsychologia* 42, no. 13 (2004): 1788–1804.

61 **326 people with a range of handedness:** Knecht et al., "Handedness and Hemispheric Language Dominance."

63 **seen in the auditory cortex:** Mari Tervaniemi and Kenneth Hugdahl, "Lateralization of Auditory-Cortex Functions," *Brain Research Reviews* 43, no. 3 (2003): 231–246.

63 **David Poeppel:** David Poeppel, "The Analysis of Speech in Different Temporal Integration Windows: Cerebral Lateralization as 'Asymmetric Sampling in Time,'" *Speech Communication* 41, no. 1 (2003): 245–255.

63 **Robert Zatorre:** Robert J. Zatorre, Pascal Belin, and Virginia B. Penhune, "Structure and Function of Auditory Cortex: Music and Speech," *Trends in Cognitive Sciences* 6, no. 1 (2002): 37–46.

64 **an astonishing rate of 2,109 drumbeats:** "Siddharth Nagarajan," Wikipedia, accessed online April 15, 2021, https://en.wikipedia.org/wiki/Siddharth_Nagarajan/.

64   **the critical structural difference:** Elkhonon Goldberg and Louis D. Costa, "Hemisphere Differences in the Acquisition and Use of Descriptive Systems," *Brain and Language* 14, no. 1 (1981): 144–173.

66   **they begin to show a stable preference:** Eliza L. Nelson, Julie M. Campbell, and George F. Michel, "Unimanual to Bimanual: Tracking the Development of Handedness from 6 to 24 Months," *Infant Behavior and Development* 36, no. 2 (2013): 181–188; and Jacqueline Fagard and Anne Marks, "Unimanual and Bimanual Tasks and the Assessment of Handedness in Toddlers," *Developmental Science* 3, no. 2 (2000): 137–147.

67   **shifts to left dominance in most:** For example, Mills et al., "Language Comprehension and Cerebral Specialization," and Margriet A. Groen et al., "Does Cerebral Lateralization Develop? A Study Using Functional Transcranial Doppler Ultrasound Assessing Lateralization for Language Production and Visuospatial Memory," *Brain and Behavior* 2, no. 3 (2012): 256–269.

67   *second* **languages rely more on the right hemisphere:** Judith Evans et al., "Differential Bilingual Laterality: Mythical Monster Found in Wales," *Brain and Language* 83, no. 2 (2002): 291–299.

67   **a shift toward left-lateralized music processing:** See, for example, Kentaro Ono et al., "The Effect of Musical Experience on Hemispheric Lateralization in Musical Feature Processing," *Neuroscience Letters* 496, no. 2 (2011): 141–145; Charles J. Limb et al., "Left Hemispheric Lateralization of Brain Activity During Passive Rhythm Perception in Musicians," *The Anatomical Record Part A: Discoveries in Molecular, Cellular, and Evolutionary Biology: An Official Publication of the American Association of Anatomists* 288, no. 4 (2006): 382–389; and Peter Vuust et al., "To Musicians, the Message Is in the Meter: Pre-Attentive Neuronal Responses to Incongruent Rhythm Are Left-Lateralized in Musicians," *Neuroimage* 24, no. 2 (2005): 560–564.

68   **motor cortex was indistinguishable:** Stefan Klöppel et al., "Nurture Versus Nature: Long-Term Impact of Forced Right-Handedness on Structure of Pericentral Cortex and Basal Ganglia," *Journal of Neuroscience* 30, no. 9 (2010): 3271–3275.

69   **have a critical evolutionary advantage:** Joseph Dien, "Looking Both Ways Through Time: The Janus Model of Lateralized Cognition," *Brain and Cognition* 67, no. 3 (2008): 292–323.

69   **left hemisphere initiates "approach" behaviors:** Helena J. V. Rutherford and Annukka K. Lindell, "Thriving and Surviving: Approach and Avoidance Motivation and Lateralization," *Emotion Review* 3, no. 3 (2011): 333–343.

71   **"French Push Bottles Up German Rear":** Ian Mayes, "Heads You Win: The Readers' Editor on the Art of the Headline Writer," *Guardian*, April 13, 2000: and "Syntactic ambiguity," Wikipedia, accessed online November 3, 2021, https://en.wikipedia.org/wiki/Syntactic_ambiguity#cite_note-13/.

72 **connections between their hemispheres surgically severed:** Michael S. Gazzaniga, Joseph E. Bogen, and Roger W. Sperry, "Some Functional Effects of Sectioning the Cerebral Commissures in Man," *Proceedings of the National Academy of Sciences* 48, no. 10 (1962): 1765–1769.

73 **Alan Alda and Gazzaniga:** "Basic Split Brain Science Primer: Alan Alda with Michael Gazzaniga," Scientific American/Frontiers Introductory Psychology Video Collection, YouTube video, uploaded by Michael Blackstone on January 5, 2017, https://www.youtube.com/watch?v=4CdmvNKwNjM/.

73 **"inferential" processes in both hemispheres:** Chantel S. Prat, Robert A. Mason, and Marcel Adam Just, "Individual Differences in the Neural Basis of Causal Inferencing," *Brain and Language* 116, no. 1 (2011): 1–13; Chantel S. Prat, Robert A. Mason, and Marcel Adam Just, "An fMRI Investigation of Analogical Mapping in Metaphor Comprehension: The Influence of Context and Individual Cognitive Capacities on Processing Demands," *Journal of Experimental Psychology: Learning, Memory, and Cognition* 38, no. 2 (2012): 282; and Chantel S. Prat, "The Brain Basis of Individual Differences in Language Comprehension Abilities," *Language and Linguistics Compass* 5, no. 9 (2011): 635–649.

73 **intact participants and callosotomy patients alike:** Matthew E. Roser et al., "Dissociating Processes Supporting Causal Perception and Causal Inference in the Brain," *Neuropsychology* 19, no. 5 (2005): 591.

75 **a tweet that went viral in January 2020:** @KylePlantEmoji tweeted "Fun fact: some people have an internal narrative and some don't. As in, some people's thoughts are like sentences they 'hear', and some people just have abstract non-verbal thoughts, and have to consciously verbalize them And most people aren't aware of the other type of person" on January 27, 2020, https://twitter.com/KylePlantEmoji/status/1221713792913965061?s=20.

76 **"I'd reach with my right [hand]":** David Wolman, "The Split Brain: A Tale of Two Halves," *Nature News* 483, no. 7389 (2012): 260.

78 **one remarkable study by Casagrande and Bertini:** Maria Casagrande and Mario Bertini, "Night-Time Right Hemisphere Superiority and Daytime Left Hemisphere Superiority: A Repatterning of Laterality Across Wake–Sleep–Wake States," *Biological Psychology* 77, no. 3 (2008): 337–342.

78 **activating your left motor cortex:** See, for example, Eddie Harmon-Jones, "Unilateral Right-Hand Contractions Cause Contralateral Alpha Power Suppression and Approach Motivational Affective Experience," *Psychophysiology* 43, no. 6 (2006): 598–603.

CHAPTER 2: MIXOLOGY

81 **the human brain has *hundreds* of different kinds:** For an incomplete but comprehensive list, see "Neurotransmitter" entry on Wikipedia, last accessed online November 4, 2021, https://en.wikipedia.org/wiki/Neurotransmitter/.

82   **at least one caffeinated beverage per day:** Diane C. Mitchell et al., "Beverage Caffeine Intakes in the US," *Food and Chemical Toxicology* 63 (2014): 136–142.

82   **increases the availability of a neurotransmitter called *dopamine*:** O. Cauli and Micaela Morelli, "Caffeine and the Dopaminergic System," *Behavioural Pharmacology* 16, no. 2 (2005): 63–77. Also see Marcello Solinas et al., "Caffeine Induces Dopamine and Glutamate Release in the Shell of the Nucleus Accumbens," *Journal of Neuroscience* 22, no. 15 (2002): 6321–6324.

83   **surgically rewiring a newborn ferret brain:** Mriganka Sur, Preston E. Garraghty, and Anna W. Roe, "Experimentally Induced Visual Projections into Auditory Thalamus and Cortex," *Science* 242, no. 4884 (1988): 1437–1441.

83   **able to take over the *function* of the seeing:** Laurie Von Melchner, Sarah L. Pallas, and Mriganka Sur, "Visual Behaviour Mediated by Retinal Projections Directed to the Auditory Pathway," *Nature* 404, no. 6780 (2000): 871–876. Also see Sandra Blakeslee, "'Rewired' Ferrets Overturn Theories of Brain Growth," *New York Times*, April 25, 2000.

84   **neural signals get crossed:** Jürgen Hänggi, Diana Wotruba, and Lutz Jäncke, "Globally Altered Structural Brain Network Topology in Grapheme-Color Synesthesia," *Journal of Neuroscience* 31, no. 15 (2011): 5816–5828.

84   **the merging of two *unrelated* streams:** Richard E. Cytowic and David Eagleman, *Wednesday Is Indigo Blue: Discovering the Brain of Synesthesia* (MIT Press, 2011). Also see David Brang and Vilayanur S. Ramachandran, "Survival of the Synesthesia Gene: Why Do People Hear Colors and Taste Words?" *PLoS Biology* 9, no. 11 (2011): e1001205.

85   **causes permanent damage to your retina:** P. A. MacFaul, "Visual Prognosis After Solar Retinopathy," *British Journal of Ophthalmology* 53, no. 8 (1969): 534.

87   **relates to hypnotic susceptibility:** Richard P. Atkinson and Helen J. Crawford, "Individual Differences in Afterimage Persistence: Relationships to Hypnotic Susceptibility and Visuospatial Skills," *American Journal of Psychology* (1992): 527–539. Also see Richard P. Atkinson, "Enhanced Afterimage Persistence in Waking and Hypnosis: High Hypnotizables Report More Enduring Afterimages," *Imagination, Cognition and Personality* 14, no. 1 (1994): 31–41.

87   **also been linked to individual differences in neurochemistry:** David J. Acunzo, David A. Oakley, and Devin B. Terhune, "The Neurochemistry of Hypnotic Suggestion," *American Journal of Clinical Hypnosis* 63, no. 4 (2021): 355–371.

88   **7.8 percent of adults in the United States:** Statistic taken from the National Institute of Mental Health's 2019 Survey, https://www.nimh.nih .gov/health/statistics/major-depression/.

91   **"Mini-Markers" personality test:** Gerard Saucier, "Mini-Markers: A Brief Version of Goldberg's Unipolar Big-Five Markers," *Journal of Personality Assessment* 63, no. 3 (1994): 506–516.

91   **related to individual differences in neurochemistry:** Richard A. Depue and Yu Fu, "Neurobiology and Neurochemistry of Temperament in Adults," in *Handbook of Temperament*, eds. M. Zentner and R. L. Shiner (New York: Guilford Press, 2012), 368–399. Also see Irina Trofimova and Trevor W. Robbins, "Temperament and Arousal Systems: A New Synthesis of Differential Psychology and Functional Neurochemistry," *Neuroscience & Biobehavioral Reviews* 64 (2016): 382–402.

92   **trick questions built in:** See, for example, Randall A. Gordon, "Social Desirability Bias: A Demonstration and Technique for Its Reduction," *Teaching of Psychology* 14, no. 1 (1987): 40–42.

93   **they largely agree:** Hans Jurgen Eysenck, "Biological Basis of Personality," *Nature* 199, no. 4898 (1963): 1031–1034. Also see Jeffrey A. Gray, "A Critique of Eysenck's Theory of Personality," in *A Model for Personality*, ed. H. J. Eysenck (Springer-Verlag, 1981), 246–276; and Gerald Matthews and Kirby Gilliland, "The Personality Theories of HJ Eysenck and JA Gray: A Comparative Review," *Personality and Individual Differences* 26, no. 4 (1999): 583–626, for review of the others.

96   **dopamine communication systems are at least partially responsible:** Richard A. Depue and Paul F. Collins, "Neurobiology of the Structure of Personality: Dopamine, Facilitation of Incentive Motivation, and Extraversion," *Behavioral and Brain Sciences* 22, no. 3 (1999): 491–517.

97   *everything* **that makes you feel good:** See, for example, Troels W. Kjaer et al., "Increased Dopamine Tone During Meditation-Induced Change of Consciousness," *Cognitive Brain Research* 13, no. 2 (2002): 255–259; and Jeffrey M. Brown, Glen R. Hanson, and Annette E. Fleckenstein, "Methamphetamine Rapidly Decreases Vesicular Dopamine Uptake," *Journal of Neurochemistry* 74, no. 5 (2000): 2221–2223.

100   **among the first to demonstrate this in the lab:** Michael X. Cohen et al., "Individual Differences in Extraversion and Dopamine Genetics Predict Neural Reward Responses," *Cognitive Brain Research* 25, no. 3 (2005): 851–861.

100   **"How does an MRI work":** Link to NIH website describing MRI technology, https://www.nibib.nih.gov/science-education/science-topics/magnetic-resonance-imaging-mri/.

103   **provided the data that closed this loop:** Luke D. Smillie et al., "Variation in DRD2 Dopamine Gene Predicts Extraverted Personality," *Neuroscience Letters* 468, no. 3 (2010): 234–237.

103   **evidence linking extraversion to dopamine:** See, for example, Luke D. Smillie et al., "Extraversion and Reward-Processing: Consolidating Evidence from an Electroencephalographic Index of Reward-Prediction-Error," *Biological Psychology* 146 (2019): 107735.

104   **people received news about unexpected rewards:** Luke D. Smillie, Andrew J. Cooper, and Alan D. Pickering, "Individual Differences in Reward–Prediction–Error: Extraversion and Feedback-Related Negativity," *Social Cognitive and Affective Neuroscience* 6, no. 5 (2011): 646–652.

104 **extraverts also tend to rate themselves as happier:** David Watson and Lee Anna Clark, "Extraversion and Its Positive Emotional Core," in *Handbook of Personality Psychology*, eds. Robert Hogan, John A. Johnson, and Stephen R. Briggs (Academic Press, 1997), 767–793. Also see William Pavot, E. D. Diener, and Frank Fujita, "Extraversion and Happiness," *Personality and Individual Differences* 11, no. 12 (1990): 1299–1306; and Michael Argyle and Luo Lu, "The Happiness of Extraverts," *Personality and Individual Differences* 11, no. 10 (1990): 1011–1017.

105 **put an electrode directly into the part of a rat's brain:** J. Olds and Peter Milner, "Positive Reinforcement Produced by Electrical Stimulation of Septal Area and Other Brain Regions in the Rat," *Comparative Physiology* 47, no. 6 (1954): 419–427.

105 **as much as five thousand times per hour:** James Olds, "Pleasure Centers in the Brain," *Scientific American* 195, no. 4 (1956): 105–117.

105 **has also been associated with obesity:** Xue Sun, Serge Luquet, and Dana M. Small, "DRD2: Bridging the Genome and Ingestive Behavior," *Trends in Cognitive Sciences* 21, no. 5 (2017): 372–384.

106 **has been related to anhedonia:** Andre Der-Avakian and Athina Markou, "The Neurobiology of Anhedonia and Other Reward-Related Deficits," *Trends in Neurosciences* 35, no. 1 (2012): 68–77.

106 **the satiety signal:** Y-Lan Boureau and Peter Dayan, "Opponency Revisited: Competition and Cooperation Between Dopamine and Serotonin," *Neuropsychopharmacology* 36, no. 1 (2011): 74–97.

106 **is created in your digestive tract:** See, for example, Jessica M. Yano et al., "Indigenous Bacteria from the Gut Microbiota Regulate Host Serotonin Biosynthesis," *Cell* 161, no. 2 (2015): 264–276.

107 **Unless you inject serotonin into their brains!:** Nuria de Pedro et al., "Inhibitory Effect of Serotonin on Feeding Behavior in Goldfish: Involvement of CRF," *Peptides* 19, no. 3 (1998): 505–511.

107 **when serotonin goes down and dopamine goes up:** Jeffrey W. Dalley and J. P. Roiser, "Dopamine, Serotonin and Impulsivity," *Neuroscience* 215 (2012): 42–58.

108 **the results are *not* as straightforward:** J. A. Schinka, R. M. Busch, and N. Robichaux-Keene, "A Meta-Analysis of the Association Between the Serotonin Transporter Gene Polymorphism (5—HTTLPR) and Trait Anxiety," *Molecular Psychiatry* 9, no. 2 (2004): 197–202.

108 **as many as 1 in 3 people don't improve:** Alessandro Serretti and Masaki Kato, "The Serotonin Transporter Gene and Effectiveness of SSRIs," *Expert Review of Neurotherapeutics* 8, no. 1 (2008): 111–120.

109 **rated themselves higher on anxiety-related personality characteristics:** Klaus-Peter Lesch et al., "Association of Anxiety-Related Traits with a Polymorphism in the Serotonin Transporter Gene Regulatory Region," *Science* 274, no. 5292 (1996): 1527–1531.

110 **results have not been consistently replicated:** J. D. Flory et al., "Neuroticism Is Not Associated with the Serotonin Transporter (5—HTTLPR) Polymorphism," *Molecular Psychiatry* 4, no. 1 (1999): 93–96.

110 **results varied depending on the specific personality measure:** Flory et al., "Neuroticism Is Not Associated," xxxii.

111 **a brain that is not responding to environmental stressors in either a typical or functional way:** Hymie Anisman and Robert M. Zacharko, "Depression as a Consequence of Inadequate Neurochemical Adaptation in Response to Stressors," *British Journal of Psychiatry* 160, no. S15 (1992): 36–43.

111 **all *typical* brain responses to stress:** Nicole Baumann and Jean-Claude Turpin, "Neurochemistry of Stress: An Overview," *Neurochemical Research* 35, no. 12 (2010): 1875–1879.

112 **genes related to serotonin reuptake:** Baldwin M. Way and Shelley E. Taylor, "The Serotonin Transporter Promoter Polymorphism Is Associated with Cortisol Response to Psychosocial Stress," *Biological Psychiatry* 67, no. 5 (2010): 487–492.

115 **to counteract the effects:** Steven E. Hyman and Eric J. Nestler, "Initiation and Adaptation: A Paradigm for Understanding Psychotropic Drug Action," *American Journal of Psychiatry* (1996).

115 **people who reduce their caffeine intake:** Laura M. Juliano and Roland R. Griffiths, "A Critical Review of Caffeine Withdrawal: Empirical Validation of Symptoms and Signs, Incidence, Severity, and Associated Features," *Psychopharmacology* 176, no. 1 (2004): 1–29.

117 **temporarily reduce the amount of serotonin communication:** W. A. Williams et al., "Effects of Acute Tryptophan Depletion on Plasma and Cerebrospinal Fluid Tryptophan and 5-Hydroxyindoleacetic Acid in Normal Volunteers," *Journal of Neurochemistry* 72, no. 4 (1999): 1641–1647.

117 **effects of tyrosine on blood pressure and anxiety levels:** J. B. Deijen and J. F. Orlebeke, "Effect of Tyrosine on Cognitive Function and Blood Pressure Under Stress," *Brain Research Bulletin* 33, no. 3 (1994): 319–323, and J. B. Deijen et al., "Tyrosine Improves Cognitive Performance and Reduces Blood Pressure in Cadets After One Week of a Combat Training Course," *Brain Research Bulletin* 48, no. 2 (1999): 203–209. But see also Lydia A. Conlay, Timothy J. Maher, and Richard J. Wurtman, "Tyrosine Increases Blood Pressure in Hypotensive Rats," *Science* 212, no. 4494 (1981): 559–560.

117 **moderate levels of aerobic exercise:** See, for example, Romain Meeusen and Kenny De Meirleir, "Exercise and Brain Neurotransmission," *Sports Medicine* 20, no. 3 (1995): 160–188.

117 **"good stress":** Saskia Heijnen et al., "Neuromodulation of Aerobic Exercise—A Review," *Frontiers in Psychology* 6 (2016): 1890.

118   **Massage therapy:** Tiffany Field et al., "Cortisol Decreases and Serotonin and Dopamine Increase Following Massage Therapy," *International Journal of Neuroscience* 115, no. 10 (2005): 1397–1413.

118   **Meditation and mindfulness practices:** Rose H. Matousek, Patricia L. Dobkin, and Jens Pruessner, "Cortisol as a Marker for Improvement in Mindfulness-Based Stress Reduction," *Complementary Therapies in Clinical Practice* 16, no. 1 (2010): 13–19; and Kenneth G. Walton et al., "Stress Reduction and Preventing Hypertension: Preliminary Support for a Psychoneuroendocrine Mechanism," *Journal of Alternative and Complementary Medicine* 1, no. 3 (1995): 263–283.

118   **deep-breathing exercises:** Valentina Perciavalle et al., "The Role of Deep Breathing on Stress," *Neurological Sciences* 38, no. 3 (2017): 451–458.

CHAPTER 3: IN SYNC

122   **a massive rhythm generator:** Gyorgy Buzsaki, *Rhythms of the Brain* (Oxford University Press, 2006).

125   **following the same principles of *network theory*:** Duncan J. Watts and Steven H. Strogatz, "Collective Dynamics of 'Small-World' Networks," *Nature* 393, no. 6684 (1998): 440–442.

126   **they are covered in myelin:** For a comprehensive overview, see R. Douglas Fields, "White Matter Matters," *Scientific American* 298, no. 3 (2008): 54–61.

127   **connects the neurons responsible for visual pattern detection:** See, for example, Brian A. Wandell, Andreas M. Rauschecker, and Jason D. Yeatman, "Learning to See Words," *Annual Review of Psychology* 63 (2012): 31–53.

127   **The "siren call" of word reading:** J. Ridley Stroop, "Studies of Interference in Serial Verbal Reactions," *Journal of Experimental Psychology* 18, no. 6 (1935): 643.

130   **can produce speech at rates of over twelve syllables per second:** Amalajobitha, "Who Is the Fastest Rapper in the World 2021?" *Freshers Live,* July 28, 2021, https://latestnews.fresherslive.com/articles/fastest-rapper-in-the-world-who-is-the-fastest-rapper-in-the-world-261359/.

131   **orchestrate the higher-frequency neural processes:** Earl K. Miller, Mikael Lundqvist, and André M. Bastos, "Working Memory 2.0," *Neuron* 100, no. 2 (2018): 463–475.

131   **why we *suck* at multitasking:** There are many different ideas about this, but here's an example of a study that implicates signal interference in multitasking performance: Menno Nijboer et al., "Single-Task fMRI Overlap Predicts Concurrent Multitasking Interference," *NeuroImage* 100 (2014): 60–74.

131   **the *attentional blink*:** Kimron L. Shapiro, Jane E. Raymond, and Karen M. Arnell, "The Attentional Blink," *Trends in Cognitive Sciences* 1, no. 8 (1997): 291–296.

132   **wide-reaching implications for how different people process information:** See, for example, Chantel S. Prat et al., "Resting-State qEEG Predicts

Rate of Second Language Learning in Adults," *Brain and Language* 157 (2016): 44–50; and Chantel S. Prat, Brianna L. Yamasaki, and Erica R. Peterson, "Individual Differences in Resting-State Brain Rhythms Uniquely Predict Second Language Learning Rate and Willingness to Communicate in Adults," *Journal of Cognitive Neuroscience* 31, no. 1 (2019): 78–94.

135   **Torrance Tests of Creative Thinking:** E. Paul Torrance, "Predictive Validity of the Torrance Tests of Creative Thinking," *Journal of Creative Behavior* (1972).

137   **borrowed from the Compound Remote Associates Test:** Edward M. Bowden and Mark Jung-Beeman, "Normative Data for 144 Compound Remote Associate Problems," *Behavior Research Methods, Instruments & Computers* 35, no. 4 (2003): 634–639.

141   **"Were you rushing or were you dragging?":** Damien Chazelle, dir., *Whiplash,* Sony Pictures Classics, 2014.

141   **measured alpha frequency and working memory capacity:** C. Richard Clark et al., "Spontaneous Alpha Peak Frequency Predicts Working Memory Performance Across the Age Span," *International Journal of Psychophysiology* 53, no. 1 (2004): 1–9.

143   **insight versus deliberate search processes:** Brian Erickson et al., "Resting-State Brain Oscillations Predict Trait-Like Cognitive Styles," *Neuropsychologia* 120 (2018): 1–8.

145   **Perhaps the best evidence:** Roberto Cecere, Geraint Rees, and Vincenzo Romei, "Individual Differences in Alpha Frequency Drive Crossmodal Illusory Perception," *Current Biology* 25, no. 2 (2015): 231–235.

147   **the rate at which they can *refresh* the bits of information:** W. Klimesch et al., "Alpha Frequency, Reaction Time, and the Speed of Processing Information," *Journal of Clinical Neurophysiology* 13, no. 6 (1996): 511–518. Also see Thomas H. Grandy et al., "Individual Alpha Peak Frequency Is Related to Latent Factors of General Cognitive Abilities," *Neuroimage* 79 (2013): 10–18, for a related discussion of alpha frequency and cognition more broadly defined.

148   **measured creativity using a task similar to the triangle test:** O. M. Bazanova and L. I. Aftanas, "Individual Measures of Electroencephalogram Alpha Activity and Non-Verbal Creativity," *Neuroscience and Behavioral Physiology* 38, no. 3 (2008): 227–235.

149   **systems that are most heritable:** C. M. Smit et al., "Genetic Variation of Individual Alpha Frequency (IAF) and Alpha Power in a Large Adolescent Twin Sample," *International Journal of Psychophysiology* 61, no. 2 (2006): 235–243.

149   **change across our life spans:** Smit et al., "Genetic Variation of IAF," xiv.

150   **the average alpha frequency will increase about 5.5 Hz:** John R. Hughes and Juan J. Cayaffa, "The EEG in Patients at Different Ages Without Organic Cerebral Disease," *Electroencephalography and Clinical Neurophysiology* 42, no. 6 (1977): 776–784.

150 **shown to influence neural synchronization in interesting ways:** See, for example, Tim Lomas, Itai Ivtzan, and Cynthia H. Y. Fu, "A Systematic Review of the Neurophysiology of Mindfulness on EEG Oscillations," *Neuroscience & Biobehavioral Reviews* 57 (2015): 401–410.

150 **three months of intensive meditation training:** Manish Saggar et al., "Intensive Training Induces Longitudinal Changes in Meditation State-Related EEG Oscillatory Activity," *Frontiers in Human Neuroscience* 6 (2012): 256.

150–51 **Not all meditation studies:** For review, see B. Rael Cahn and John Polich, "Meditation States and Traits: EEG, ERP, and Neuroimaging Studies," *Psychological Bulletin* 132, no. 2 (2006): 180.

151 **gaming can increase people's peak alpha frequency ranges:** Cameron Sheikholeslami et al., "A High Resolution EEG Study of Dynamic Brain Activity During Video Game Play," in *29th Annual International Conference of the IEEE Engineering in Medicine and Biology Society* (IEEE, 2007): 2489–2491.

151 **consuming 250 milligrams of caffeine:** Robert J. Barry et al., "Caffeine Effects on Resting-State Arousal," *Clinical Neurophysiology* 116, no. 11 (2005): 2693–2700.

CHAPTER 4: FOCUS

158 **a highly overqualified joystick:** Rajesh P. N. Rao et al., "A Direct Brain-to-Brain Interface in Humans," *PLOS ONE* 9, no. 11 (2014): e111332.

158 **the original footage on YouTube:** "Direct Brain-to-Brain Communication in Humans: A Pilot Study," YouTube video, uploaded by uwneuralsystems, August 26, 2013, https://www.youtube.com/watch?v=rNRDc714WSI/.

158 **documentary *I Am Human*:** Elena Gaby and Taryn Southern, dirs., *I Am Human*, Futurism Studios, March 3, 2020.

159 **"Pizzagate":** "Pizzagate conspiracy theory," Wikipedia, accessed online November 5, 2021, https://en.wikipedia.org/wiki/Pizzagate_conspiracy_theory/.

162 **The most common type of neglect:** S. P. Stone, P. W. Halligan, and R. J. Greenwood, "The Incidence of Neglect Phenomena and Related Disorders in Patients with an Acute Right or Left Hemisphere Stroke," *Age and Ageing* 22, no. 1 (1993): 46–52.

162 **leaving the left half blank:** See Andrew Parton, Paresh Malhotra, and Masud Husain, "Hemispatial Neglect," *Journal of Neurology, Neurosurgery & Psychiatry* 75, no. 1 (2004): 13–21, for fascinating examples of these assessments.

163 **how likely someone is to seek out:** B. Gialanella and F. Mattioli, "Anosognosia and Extrapersonal Neglect as Predictors of Functional Recovery Following Right Hemisphere Stroke," *Neuropsychological Rehabilitation* 2, no. 3 (1992): 169–178.

163   **damage to the left hemisphere rarely:** Elisabeth Becker and Hans-Otto Karnath, "Incidence of Visual Extinction After Left Versus Right Hemisphere Stroke," *Stroke* 38, no. 12 (2007): 3172–3174.

163   **dramatic differences in attentional deficits:** Guido Gainotti, "Lateralization of Brain Mechanisms Underlying Automatic and Controlled Forms of Spatial Orienting of Attention," *Neuroscience & Biobehavioral Reviews* 20, no. 4 (1996): 617–622.

167   **Most *typical* left-hemisphere-dominant people:** See control groups in Naren Prahlada Rao et al., "Lateralisation Abnormalities in Obsessive–Compulsive Disorder: A Line Bisection Study," *Acta Neuropsychiatrica* 27, no. 4 (2015): 242–247; and Karen E. Waldie and Markus Hausmann, "Right Fronto-Parietal Dysfunction in Children with ADHD and Developmental Dyslexia as Determined by Line Bisection Judgements," *Neuropsychologia* 48, no. 12 (2010): 3650–3656.

168   **regularly bisect lines to the right of their true center:** Waldie and Hausmann, "Right Fronto-Parietal Dysfunction."

168   **diagnosed more frequently in non-right-handed individuals:** Eunice N. Simões, Ana Lucia Novais Carvalho, and Sergio L. Schmidt, "What Does Handedness Reveal About ADHD? An Analysis Based on CPT Performance," *Research in Developmental Disabilities* 65 (2017): 46–56; and Evgenia Nastou, Sebastian Ocklenburg, and Marietta Papadatou-Pastou, "Handedness in ADHD: Meta-Analyses," *PsyArXiv* (2020), https://psyarxiv.com/zyrvg/.

171   **using a tactile discrimination task:** Saskia Haegens, Barbara F. Händel, and Ole Jensen, "Top-Down Controlled Alpha Band Activity in Somatosensory Areas Determines Behavioral Performance in a Discrimination Task," *Journal of Neuroscience* 31, no. 14 (2011): 5197–5204.

171   **how people perform on the Stroop Task:** Rebecca J. Compton et al., "Cognitive Control in the Intertrial Interval: Evidence from EEG Alpha Power," *Psychophysiology* 48, no. 5 (2011): 583–590.

172   **alpha in the left hemisphere:** Brian Erickson et al., "Resting-State Brain Oscillations Predict Trait-Like Cognitive Styles," *Neuropsychologia* 120 (2018): 1–8.

172   **"turtles all the way down":** For explanation see "Turtles all the way down," Wikipedia, accessed on 04/02/2020, https://en.wikipedia.org/wiki/Turtles_all_the_way_down/.

173   **Robert Sapolsky's discussion:** Robert M. Sapolsky, *Behave: The Biology of Humans at Our Best and Worst* (Penguin, 2017).

176   **its legal name is "Conditional Routing Model":** Andrea Stocco, Christian Lebiere, and John R. Anderson, "Conditional Routing of Information to the Cortex: A Model of the Basal Ganglia's Role in Cognitive Coordination," *Psychological Review* 117, no. 2 (2010): 541.

176 **brain responses of people with different working memory capacities:** Chantel S. Prat and Marcel Adam Just, "Exploring the Neural Dynamics Underpinning Individual Differences in Sentence Comprehension," *Cerebral Cortex* 21, no. 8 (2011): 1747–1760.

177 **whether bilingualism has any effect:** See, for example, Andrea Stocco and Chantel S. Prat, "Bilingualism Trains Specific Brain Circuits Involved in Flexible Rule Selection and Application," *Brain and Language* 137 (2014): 50–61; and A. Stocco et al., "Bilingual Brain Training: A Neurobiological Framework of How Bilingual Experience Improves Executive Function," *International Journal of Bilingualism* 18, no. 1 (2014): 67–92.

177 **the neural basis of autism spectrum disorder:** See the original paper: Rajesh K. Kana, Lauren E. Libero, and Marie S. Moore, "Disrupted Cortical Connectivity Theory as an Explanatory Model for Autism Spectrum Disorders," *Physics of Life Reviews* 8, no. 4 (2011): 410–437; and our commentary: Chantel S. Prat and Andrea Stocco, "Information Routing in the Basal Ganglia: Highways to Abnormal Connectivity in Autism?: Comment on 'Disrupted Cortical Connectivity Theory as an Explanatory Model for Autism Spectrum Disorders' by Kana et al.," *Physics of Life Reviews* 9, no. 1 (2012): 1.

179 **we analyzed functional MRI data:** Chantel S. Prat et al., "Basal Ganglia Impairments in Autism Spectrum Disorder Are Related to Abnormal Signal Gating to Prefrontal Cortex," *Neuropsychologia* 91 (2016): 268–281.

CHAPTER 5: ADAPT

184 **underwent a "cognitive revolution":** See, for example, John Medina, *Brain Rules: 12 Principles for Surviving and Thriving at Work, Home, and School* (Seattle: Pear Press, 2011).

184 **increased following periods of extreme weather instability:** Jessica Ash and Gordon G. Gallup, "Paleoclimatic Variation and Brain Expansion During Human Evolution," *Human Nature* 18, no. 2 (2007): 109–124.

185 **poetically described what it must be like:** William James, *Principles of Psychology* (1863).

189 **experience leaves its mark:** For a recent review, see Arun Asok et al., "Molecular Mechanisms of the Memory Trace," *Trends in Neurosciences* 42, no. 1 (2019): 14–22.

189 **process called *Hebbian learning*:** There is an accessible write-up of this on the Decision Lab's website: https://thedecisionlab.com/reference-guide/neuroscience/hebbian-learning/.

190 **"Neurons that fire together, wire together":** This phrase was credited to Hebb himself (1949) in an article written on the SuperCamp website, but I could not verify the source elsewhere: https://www.supercamp.com/what-does-neurons-that-fire-together-wire-together-mean/.

192 **when infants are immersed in their native languages:** Patricia K. Kuhl, "The Development of Speech and Language," *Mechanistic Relationships Between Development and Learning* (1998): 53–73.

192 **"citizens of the world":** Patricia Kuhl, "The Linguistic Genius of Babies," TED Talk, uploaded to YouTube on February 18, 2011, https://www.you tube.com/watch?v=G2XBIkHW954/.

194 **"How to do a pullover":** "How to do a Pullover on Bars," uploaded by TC2 on December 2, 2014, https://www.youtube.com/watch?v=DzW1 TnJChD0/.

195 **improvements after mental practice:** See, for example, Richard M. Suinn, "Mental Practice in Sport Psychology: Where Have We Been, Where Do We Go?" *Clinical Psychology: Science and Practice* 4, no. 3 (1997): 189–207; Lars Nyberg et al., "Learning by Doing Versus Learning by Thinking: An fMRI Study of Motor and Mental Training," *Neuropsychologia* 44, no. 5 (2006): 711–717; and Carl-Johan Olsson, Bert Jonsson, and Lars Nyberg, "Learning by Doing and Learning by Thinking: An fMRI Study of Combining Motor and Mental Training," *Frontiers in Human Neuroscience* 2 (2008): 5.

196 **from the Language Experience and Proficiency Questionnaire (LEAP-Q):** Viorica Marian, Henrike K. Blumenfeld, and Margarita Kaushanskaya, "Language Experience and Proficiency Questionnaire (LEAP-Q)" (2018).

199 **four aspects of language experience:** A. Stocco et al., "Bilingual Brain Training: A Neurobiological Framework of How Bilingual Experience Improves Executive Function," *International Journal of Bilingualism* 18, no. 1 (2014): 67–92; Brianna L. Yamasaki, Andrea Stocco, and Chantel S. Prat, "Relating Individual Differences in Bilingual Language Experiences to Executive Attention," *Language, Cognition and Neuroscience* 33, no. 9 (2018): 1128–1151; Kinsey Bice et al., "Bilingual Language Experience Shapes Resting-State Brain Rhythms," *Neurobiology of Language* 1, no. 3 (2020): 288–318; and Brianna L. Yamasaki et al., "Effects of Bilingual Language Experience on Basal Ganglia Computations: A Dynamic Causal Modeling Test of the Conditional Routing Model," *Brain and Language* 197 (2019): 104665.

200 **between 20,000 and 35,000 English words:** Lexical facts, *The Economist*, May 29, 2013, https://www.economist.com/johnson/2013/05/29/lexical -facts based on data collected from testyourvocab.com/.

201 **the French documentary *Babies*:** Thomas Balmès, dir., *Babies*, Focus Features, April 14, 2010.

201 **raised in very specific visual environments:** See, for example, Helmut V. B. Hirsch and D. N. Spinelli, "Visual Experience Modifies Distribution of Horizontally and Vertically Oriented Receptive Fields in Cats," *Science* 168, no. 3933 (1970): 869–871; Helmut V. B. Hirsch and D. N. Spinelli, "Modification of the Distribution of Receptive Field Orientation in Cats by Selective Visual Exposure During Development," *Experimental Brain Research* 12, no. 5 (1971): 509–527; and N. W. Daw and H. J. Wyatt, "Kittens Reared in a Unidirectional Environment: Evidence for a Critical Period," *Journal of Physiology* 257, no. 1 (1976): 155–170.

204 **Wallisch tested one interesting hypothesis:** Pascal Wallisch, "Illumination Assumptions Account for Individual Differences in the Perceptual Interpretation of a Profoundly Ambiguous Stimulus in the Color Domain: 'The Dress,'" *Journal of Vision* 17, no. 4 (2017): 5.

205 **The effect was first demonstrated:** B. Keith Payne, "Prejudice and Perception: The Role of Automatic and Controlled Processes in Misperceiving a Weapon," *Journal of Personality and Social Psychology* 81, no. 2 (2001): 181.

206 **chilling, real-world implications:** B. Keith Payne, "Weapon Bias: Split-Second Decisions and Unintended Stereotyping," *Current Directions in Psychological Science* 15, no. 6 (2006): 287–291.

207 **Malcolm Gladwell covers this:** Malcolm Gladwell, *Blink: The Power of Thinking Without Thinking* (Little, Brown, 2006).

209 **when a bilingual wants to speak:** See, for example, Judith F. Kroll et al., "Language Selection in Bilingual Speech: Evidence for Inhibitory Processes," *Acta Psychologica* 128, no. 3 (2008): 416–430.

210 **quicker than monolinguals when executing new *mathematical* tasks:** Andrea Stocco and Chantel S. Prat, "Bilingualism Trains Specific Brain Circuits Involved in Flexible Rule Selection and Application," *Brain and Language* 137 (2014): 50–61.

210 **evidence of such increased control:** Bice, Yamasaki, and Prat, "Bilingual Language Experience."

CHAPTER 6: NAVIGATE

214 **Oprah talking about the influence it had:** Oprah Winfrey, *Oprah's Life Class*, first aired October 19, 2011, https://www.oprah.com/oprahs-lifeclass/the-powerful-lesson-maya-angelou-taught-oprah-video/.

215 **"Awareness is the greatest agent for change":** Eckhart Tolle, *A New Earth: Awakening to Your Life's Purpose* (Penguin, 2006).

215 **"It's not a case of":** Daniel Kahneman, on *Thinking Fast and Slow* (2011) as cited in Ariella S. Kristal and Laurie R. Santos, "GI Joe Phenomena: Understanding the Limits of Metacognitive Awareness on Debiasing," Harvard Business School Working Paper, 2021.

220 **the "dance lesson":** Andy Tennant, dir., *Hitch*, Sony Pictures, February 11, 2005. You can find this scene on YouTube: "Hitch (6/8) Movie CLIP-Dance Lessons (2005) HD," uploaded by Movieclips on October 6, 2012.

222 **to figure out the relative strength:** Michael J. Frank, Lauren C. Seeberger, and Randall C. O'Reilly, "By Carrot or by Stick: Cognitive Reinforcement Learning in Parkinsonism," *Science* 306, no. 5703 (2004): 1940–1943.

225 **Raven's Advanced Progressive Matrices:** J. Raven, J. C. Raven, and J. H. Court, *Manual for Raven's Advanced Progressive Matrices* (Oxford Psychologists Press, 1998).

227 **the stronger people's Avoid learning pathways are:** Andrea Stocco, Chantel S. Prat, and Lauren K. Graham, "Individual Differences in Reward-

Based Learning Predict Fluid Reasoning Abilities," *Cognitive Science* 45, no. 2 (2021): e12941.

233 **"tip-of-the-tongue" moments:** Roger Brown and David McNeill, "The 'Tip of the Tongue' Phenomenon," *Journal of Verbal Learning and Verbal Behavior* 5, no. 4 (1966): 325–337.

233 **they get worse with age:** Meredith A. Shafto et al., "On the Tip-of-the-Tongue: Neural Correlates of Increased Word-Finding Failures in Normal Aging," *Journal of Cognitive Neuroscience* 19, no. 12 (2007): 2060–2070; and Christopher J. Schmank and Lori E. James, "Adults of All Ages Experience Increased Tip-of-the-Tongue States Under Ostensible Evaluative Observation," *Aging, Neuropsychology, and Cognition* 27, no. 4 (2020): 517–531.

235 **If you want to see something fun:** "1,000,000 Dominoes Falling Is Oddly SATISFYING," YouTube, uploaded by Hevesh5 on December 2, 2017, https://www.youtube.com/watch?v=DQQN_79QrDY/.

237 **The ability to re-create a pattern:** Kazumasa Z. Tanaka et al., "Cortical Representations Are Reinstated by the Hippocampus During Memory Retrieval," *Neuron* 84, no. 2 (2014): 347–354.

237 **part of the brain that supports spatial navigation:** Georg F. Striedter, "Evolution of the Hippocampus in Reptiles and Birds," *Journal of Comparative Neurology* 524, no. 3 (2016): 496–517.

237 **appropriately named *place cells*:** Laura Lee Colgin, "Five Decades of Hippocampal Place Cells and EEG Rhythms in Behaving Rats," *Journal of Neuroscience* 40, no. 1 (2020): 54–60.

237 **creating *meaning maps* based on our experiences:** Jacob L. S. Bellmund et al., "Navigating Cognition: Spatial Codes for Human Thinking," *Science* 362, no. 6415 (2018).

239 **the responses of over 6,000 English-speakers:** Douglas L. Nelson, Cathy L. McEvoy, and Thomas A. Schreiber, "The University of South Florida Word Association, Rhyme, and Word Fragment Norms," http://w3.usf.edu/FreeAssociation/.

240 **there are fundamentally different *types* of maps:** Charan Ranganath and Maureen Ritchey, "Two Cortical Systems for Memory-Guided Behaviour," *Nature Reviews Neuroscience* 13, no. 10 (2012): 713–726.

240 **either semantic knowledge or spatial information:** Mladen Sormaz et al., "Knowing What from Where: Hippocampal Connectivity with Temporoparietal Cortex at Rest Is Linked to Individual Differences in Semantic and Topographic Memory," *Neuroimage* 152 (2017): 400–410.

244 **multivoxel pattern analysis:** James V. Haxby et al., "Distributed and Overlapping Representations of Faces and Objects in Ventral Temporal Cortex," *Science* 293, no. 5539 (2001): 2425–2430.

244 **showed participants ten different line drawings:** Svetlana V. Shinkareva et al., "Using fMRI Brain Activation to Identify Cognitive States Associated with Perception of Tools and Dwellings," *PloS one* 3, no. 1 (2008): e1394.

246 **investigated patterns of activation in sixty concrete objects:** Marcel Adam Just et al., "A Neurosemantic Theory of Concrete Noun Representation Based on the Underlying Brain Codes," *PloS one* 5, no. 1 (2010): e8622.

248 **demonstrated another chilling implication:** Marcel Adam Just et al., "Machine Learning of Neural Representations of Suicide and Emotion Concepts Identifies Suicidal Youth," *Nature Human Behaviour* 1, no. 12 (2017): 911–919.

249 **systematic differences in what people pay attention to:** Katherine L. Alfred, Megan E. Hillis, and David J. M. Kraemer, "Individual Differences in the Neural Localization of Relational Networks of Semantic Concepts," *Journal of Cognitive Neuroscience* 33, no. 3 (2021): 390–401.

CHAPTER 7: EXPLORE

257 **some species of jellyfish are capable:** For a fun read with a video, see Emily Osterloff, "Immortal Jellyfish: The Secret to Cheating Death," *What on Earth?* Natural History Museum, viewed November 9, 2021, https://www.nhm.ac.uk/discover/immortal-jellyfish-secret-to-cheating-death.html/.

259 **spontaneous pointing gestures in infants and toddlers:** Kelsey Lucca and Makeba Parramore Wilbourn, "Communicating to Learn: Infants' Pointing Gestures Result in Optimal Learning," *Child Development* 89, no. 3 (2018): 941–960; and Kelsey Lucca and Makeba Parramore Wilbourn, "The What and the How: Information-Seeking Pointing Gestures Facilitate Learning Labels and Functions," *Journal of Experimental Child Psychology* 178 (2019): 417–436.

261 **depends on *what you already know* about the world:** M. J. Gruber and C. Ranganath, "How Curiosity Enhances Hippocampus-Dependent Memory: The Prediction, Appraisal, Curiosity, and Exploration (PACE) Framework," *Trends in Cognitive Sciences* 23, no. 12 (2019): 1014–1025.

261 **"If you're having a bad day":** IFunny.co, https://ifunny.co/picture/when-you-re-having-a-bad-day-just-look-at-CvV1MzAk4/.

262 **"neither know nor think that I know":** Various translations of the Socratic paradox include "The wisest man admits that he knows nothing" or "I know that I know nothing." Though they have been attributed to Socrates, the only written account is in Plato's description of his mentor in *Apology*. This is discussed in more detail in: Gail Fine, "Does Socrates Claim to Know That He Knows Nothing?" *Oxford Studies in Ancient Philosophy* 35 (2008): 49–88.

263 **"ALL humans by nature desire to know":** The complete quote from Aristotle's *Metaphysics* says, "ALL humans by nature desire to know. An indication of this is the delight we take in our senses; for even apart from their usefulness they are loved for themselves; and above all others the sense of sight. For not only with a view to action, but even when we are not going to do anything, we prefer seeing (one might say) to everything else. The reason is that this, most of all senses, makes us know and brings to light

many differences between things." This was written in the fourth century BC and first mass-produced in 1924 by W. D. Ross.

263 **I've borrowed items:** Frank D. Naylor, "A State-Trait Curiosity Inventory," *Australian Psychologist* 16, no. 2 (1981): 172–183; and Jordan A. Litman and Charles D. Spielberger, "Measuring Epistemic Curiosity and Its Diversive and Specific Components," *Journal of Personality Assessment* 80, no. 1 (2003): 75–86.

265 **"I have no special talents":** The original "Ich habe keine besondere Begabung, sondern bin nur leidenschaftlich neugierig" appeared in a letter he wrote to Carl Seelig on March 11, 1952, *Einstein Archives 39–013.*

266 **Einstein's brain has been photographed:** D. Falk, "New Information About Albert Einstein's Brain," *Frontiers in Evolutionary Neuroscience* 1, 3 (2009); D. Falk, F. E. Lepore, and A. Noe, "The Cerebral Cortex of Albert Einstein: A Description and Preliminary Analysis of Unpublished Photographs," *Brain* 136, no. 4 (2013): 1304–1327; and W. Men et al., "The Corpus Callosum of Albert Einstein's Brain: Another Clue to His High Intelligence?" *Brain* 137, no. 4 (2014): e268–e268.

267 **observed in skilled violinists:** Peter Schwenkreis et al., "Assessment of Sensorimotor Cortical Representation Asymmetries and Motor Skills in Violin Players," *European Journal of Neuroscience* 26, no. 11 (2007): 3291–3302.

267 **a nice summary of this work:** Ashvanti Valji, "Individual Differences in Structural-Functional Brain Connections Underlying Curiosity" (PhD diss., Cardiff University, 2020).

267 **the functions of the ATL are still debated:** See, for example, Michael F. Bonner and Amy R. Price, "Where Is the Anterior Temporal Lobe and What Does It Do?" *Journal of Neuroscience* 33, no. 10 (2013): 4213–4215.

267 **it forms a hub in the brain:** Giovanna Mollo et al., "Oscillatory Dynamics Supporting Semantic Cognition: MEG Evidence for the Contribution of the Anterior Temporal Lobe Hub and Modality-Specific Spokes," *PloS One* 12, no. 1 (2017): e0169269.

268 **Diffusion imaging tracks the movement:** Wikipedia does a decent job describing this. "Diffusion MRI," Wikipedia, accessed on November 9, 2021, https://en.wikipedia.org/wiki/Diffusion_MRI/.

268 **between trait levels of epistemic curiosity and the organization of the ILF:** Ashvanti Valji et al., "Curious Connections: White Matter Pathways Supporting Individual Differences in Epistemic and Perceptual Curiosity," bioRxiv.org (2019): 642165.

270 **using a trivia experiment like I described:** Min Jeong Kang et al., "The Wick in the Candle of Learning: Epistemic Curiosity Activates Reward Circuitry and Enhances Memory," *Psychological Science* 20, no. 8 (2009): 963–973.

271 **after young adults spend six days learning Morse code:** Lara Schlaffke et al., "Learning Morse Code Alters Microstructural Properties in the

Inferior Longitudinal Fasciculus: A DTI Study," *Frontiers in Human Neuroscience* 11 (2017): 383.

273   **dopamine will be released at the cue:** Wolfram Schultz, Peter Dayan, and P. Read Montague, "A Neural Substrate of Prediction and Reward," *Science* 275, no. 5306 (1997): 1593–1599.

273   **"The Wick in the Candle of Learning":** Kang et al., "The Wick in the Candle," xvii.

274   **interested in the element of surprise:** Romain Ligneul, Martial Mermillod, and Tiffany Morisseau, "From Relief to Surprise: Dual Control of Epistemic Curiosity in the Human Brain," *NeuroImage* 181 (2018): 490–500.

276   **explored the influence of curiosity on learning:** Matthias J. Gruber, Bernard D. Gelman, and Charan Ranganath, "States of Curiosity Modulate Hippocampus-Dependent Learning via the Dopaminergic Circuit," *Neuron* 84, no. 2 (2014): 486–496.

279   **the most "metal" curiosity experiment of all time:** Johnny King L. Lau et al., "Shared Striatal Activity in Decisions to Satisfy Curiosity and Hunger at the Risk of Electric Shocks," *Nature Human Behaviour* 4, no. 5 (2020): 531–543.

285   **the relation between personal values:** Jay J. Van Bavel and Andrea Pereira, "The Partisan Brain: An Identity-Based Model of Political Belief," *Trends in Cognitive Sciences* 22, no. 3 (2018): 213–224.

CHAPTER 8: CONNECT

287   **real-world examples of misunderstandings:** Malcolm Gladwell, *Talking to Strangers: What We Should Know About the People We Don't Know* (Penguin UK, 2019).

288   **"When the snow falls and the white winds blow":** Fellow *Game of Thrones* fans will know that this was first written as a conversation between Ned Stark and Arya in George R. R. Martin, *A Game of Thrones* (A Song of Ice and Fire, Book 1) (Spectra, 1996); however, Sansa Stark said it much later in the TV series.

288   **the power of touch has been shown to help:** See, for example, Jacqueline M. McGrath, "Touch and Massage in the Newborn Period: Effects on Biomarkers and Brain Development," *Journal of Perinatal & Neonatal Nursing* 23, no. 4 (2009): 304–306.

288   **buffer the health effects of chronic illnesses like AIDS:** Jane Leserman et al., "Progression to AIDS: The Effects of Stress, Depressive Symptoms, and Social Support," *Psychosomatic Medicine* 61, no. 3 (1999): 397–406.

288   **the results of a recent meta-analysis:** Julianne Holt-Lunstad and Timothy B. Smith, "Social Relationships and Mortality," *Social and Personality Psychology Compass* 6, no. 1 (2012): 41–53.

289   **the coachable ingredients of intimate interpersonal relationships:** Jonathan W. Kanter et al., "An Integrative Contextual Behavioral Model of Intimate Relations," *Journal of Contextual Behavioral Science* (2020).

291   **one of my favorite scenes from *Westworld*:** *Westworld*, season 1, episode 6, dir. Frederick E. O. Toye, HBO, original air date November 6, 2016. See "[Westworld] Maeve 'No one knows what I'm thinking,'" YouTube, uploaded November 7, 2016, by Westworld Best Scenes, https://www.you tube.com/watch?v=qVdlnH81ON0/.

293   **an article written by philosopher Mark White:** Mark D. White, "What It Means to Know Someone," *Psychology Today*, December 2010, https://www.psychologytoday.com/us/blog/maybe-its-just-me/201012/what-it-means-know-someone/.

294   **a philosophical article on the topic written by David Matheson:** David Matheson, "Knowing Persons," *Dialogue* 49, no. 3 (2010): 435–53.

295   **one of the most frequently used measures of mind-modeling:** Simon Baron-Cohen et al., "The 'Reading the Mind in the Eyes' Test Revised Version: A Study with Normal Adults, and Adults with Asperger Syndrome or High-Functioning Autism," *Journal of Child Psychology and Psychiatry* 42, no. 2 (2001): 241–251.

297   **when your brain watches someone else performing an action:** Giacomo Rizzolatti, "The Mirror Neuron System and Its Function in Humans," *Anatomy and Embryology* 210, no. 5–6 (2005): 419–421.

298   **started by creating a social network:** Carolyn Parkinson, Adam M. Kleinbaum, and Thalia Wheatley, "Similar Neural Responses Predict Friendship," *Nature Communications* 9, no. 1 (2018): 1–14.

299   **the magnetic effect of having similar brain functioning:** Ryan Hyon et al., "Similarity in Functional Brain Connectivity at Rest Predicts Interpersonal Closeness in the Social Network of an Entire Village," *Proceedings of the National Academy of Sciences* 117, no. 52 (2020): 33149–33160.

300   **very young children don't even seem *aware*:** Janet W. Astington, Paul L. Harris, and David R. Olson, eds., *Developing Theories of Mind* (CUP Archive, 1988).

300   **some of the tests used to measure perspective-taking:** See Paul Bloom and Tim P. German, "Two Reasons to Abandon the False Belief Task as a Test of Theory of Mind," *Cognition* 77, no. 1 (2000): B25–B31; and Lynn S. Liben, "Perspective-Taking Skills in Young Children: Seeing the World Through Rose-Colored Glasses," *Developmental Psychology* 14, no. 1 (1978): 87.

300   **the extent to which this translates:** Sara M. Schaafsma et al., "Deconstructing and Reconstructing Theory of Mind," *Trends in Cognitive Sciences* 19, no. 2 (2015): 65–72.

301   **understanding how someone *feels*:** Juli Stietz et al., "Dissociating Empathy from Perspective-Taking: Evidence from Intra- and Inter-Individual Differences Research," *Frontiers in Psychiatry* 10 (2019): 126.

302   **taking the perspective of another person requires *inhibiting*:** For review, see Josef Perner and Birgit Lang, "Development of Theory of Mind

and Executive Control," *Trends in Cognitive Sciences* 3, no. 9 (1999): 337–344.

302 **measured both inhibitory control and mind modeling:** Stephanie M. Carlson and Louis J. Moses, "Individual Differences in Inhibitory Control and Children's Theory of Mind," *Child Development* 72, no. 4 (2001): 1032–1053.

303 *over 1,000 pairs* **of five-year-old twins:** Claire Hughes et al., "Origins of Individual Differences in Theory of Mind: From Nature to Nurture?" *Child Development* 76, no. 2 (2005): 356–370.

303 **"Individual Differences in Executive Functions":** Naomi P. Friedman et al., "Individual Differences in Executive Functions Are Almost Entirely Genetic in Origin," *Journal of Experimental Psychology: General* 137, no. 2 (2008): 201.

305 **the answer is probably language:** Jennifer M. Jenkins and Janet Wilde Astington, "Theory of Mind and Social Behavior: Causal Models Tested in a Longitudinal Study," *Merrill-Palmer Quarterly* 46, no. 2 (2000): 203–220.

305 **Meins first described the recursive idea:** Elizabeth Meins et al., "Rethinking Maternal Sensitivity: Mothers' Comments on Infants' Mental Processes Predict Security of Attachment at 12 Months," *Journal of Child Psychology and Psychiatry* 42, no. 5 (2001): 637–648.

305 **performed better on False Belief tests:** Elizabeth Meins et al., "Maternal Mind-Mindedness and Attachment Security as Predictors of Theory of Mind Understanding," *Child Development* 73, no. 6 (2002): 1715–1726.

306 **how likely infants were to take the "Yelp-style" recommendations:** Victoria Leong et al., "Mother-Infant Interpersonal Neural Connectivity Predicts Infants' Social Learning," *PsyArXiv* (2019), https://doi.org/10.31234/osf.io/gueaq.

308 **To get us to put in the *work* required:** See Gerald Gimpl and Falk Fahrenholz, "The Oxytocin Receptor System: Structure, Function, and Regulation," *Physiological Reviews* 81, no. 2 (2001): 629–683; and Tiffany M. Love, "Oxytocin, Motivation and the Role of Dopamine," *Pharmacology Biochemistry and Behavior* 119 (2014): 49–60.

308 **when oxytocin is on board:** Markus Heinrichs and Gregor Domes, "Neuropeptides and Social Behaviour: Effects of Oxytocin and Vasopressin in Humans," *Progress in Brain Research* 170 (2008): 337–350.

308 **Becoming a parent is among the most salient:** Naomi Scatliffe et al., "Oxytocin and Early Parent-Infant Interactions: A Systematic Review," *International Journal of Nursing Sciences* 6, no. 4 (2019): 445–453; and Ilanit Gordon et al., "Oxytocin and the Development of Parenting in Humans," *Biological Psychiatry* 68, no. 4 (2010): 377–382.

308 **oxytocin levels continue to rise in parents:** Gordon et al., "Oxytocin and the Development of Parenting."

309 **oxytocin's role in this early lamb bonding process:** Raymond Nowak et al., "Neonatal Suckling, Oxytocin, and Early Infant Attachment to the Mother," *Frontiers in Endocrinology* 11 (2021).

310 **skin-to-skin contact with either parent:** Dorothy Vittner et al., "Increase in Oxytocin from Skin-to-Skin Contact Enhances Development of Parent–Infant Relationship," *Biological Research for Nursing* 20, no. 1 (2018): 54–62.

310 **looking for the biological basis of monogamy:** Thomas R. Insel and Lawrence E. Shapiro, "Oxytocin Receptor Distribution Reflects Social Organization in Monogamous and Polygamous Voles," *Proceedings of the National Academy of Sciences* 89, no. 13 (1992): 5981–5985.

311 **Oxytocin delivered to prairie voles:** Jessie R. Williams et al., "Oxytocin Administered Centrally Facilitates Formation of a Partner Preference in Female Prairie Voles (*Microtus ochrogaster*)," *Journal of Neuroendocrinology* 6, no. 3 (1994): 247–250.

311 **a drug that blocked oxytocin binding:** T. R. Insel et al., "Oxytocin and the Molecular Basis of Monogamy," *Advances in Experimental Medicine and Biology* 395 (1995): 227–234.

311 **gave oxytocin to human participants and measured its effects:** Dirk Scheele et al., "Oxytocin Enhances Brain Reward System Responses in Men Viewing the Face of Their Female Partner," *Proceedings of the National Academy of Sciences* 110, no. 50 (2013): 20308–20313.

312 **the "real world" effects of oxytocin administration:** Dirk Scheele et al., "Oxytocin Modulates Social Distance Between Males and Females," *Journal of Neuroscience* 32, no. 46 (2012): 16074–16079.

313 **a mechanism for how oxytocin influences social bonding:** Simone G. Shamay-Tsoory and Ahmad Abu-Akel, "The Social Salience Hypothesis of Oxytocin," *Biological Psychiatry* 79, no. 3 (2016): 194–202.

314 **correlated with improved mind-modeling abilities:** See, for example, Sofia I. Cardenas et al., "Theory of Mind Processing in Expectant Fathers: Associations with Prenatal Oxytocin and Parental Attunement," *Developmental Psychobiology* (2021).

314 **males who were administered oxytocin performed better:** Gregor Domes et al., "Oxytocin Improves 'Mind-Reading' in Humans," *Biological Psychiatry* 61, no. 6 (2007): 731–733.

314 **did not find any improvements at the group level:** Sina Radke and Ellen R. A. de Bruijn, "Does Oxytocin Affect Mind-Reading? A Replication Study," *Psychoneuroendocrinology* 60 (2015): 75–81.

314 **the recognition of a handful of amygdala-related emotions:** Jenni Leppanen et al., "Meta-Analysis of the Effects of Intranasal Oxytocin on Interpretation and Expression of Emotions," *Neuroscience & Biobehavioral Reviews* 78 (2017): 125–144.

314 **oxytocin *doesn't do the same thing in everyone*:** Jennifer A. Bartz et al., "Social Effects of Oxytocin in Humans: Context and Person Matter," *Trends in Cognitive Sciences* 15, no. 7 (2011): 301–309.

315 **increased their ethnocentric, or in-group, biases:** Carsten K. W. De Dreu et al., "Oxytocin Promotes Human Ethnocentrism," *Proceedings of the National Academy of Sciences* 108, no. 4 (2011): 1262–1266.

315 **increased their sensitivity to painful expressions:** F. Sheng et al., "Oxytocin Modulates the Racial Bias in Neural Responses to Others' Suffering," *Biological Psychology* 92, no. 2 (2013): 380–386.

315 **groups that were *allegedly* based on their art preferences:** Michaela Pfundmair et al., "Oxytocin Promotes Attention to Social Cues Regardless of Group Membership," *Hormones and Behavior* 90 (2017): 136–140.

316 **should you decide to redefine what counts as your pack:** Lauren Powell et al., "The Physiological Function of Oxytocin in Humans and Its Acute Response to Human-Dog Interactions: A Review of the Literature," *Journal of Veterinary Behavior* 30 (2019): 25–32.

318 **defined as their "collective intelligence":** Anita Williams Woolley et al., "Evidence for a Collective Intelligence Factor in the Performance of Human Groups," *Science* 330, no. 6004 (2010): 686–688.

319 **predicts team performance on classroom projects:** Lisa Bender et al., Social Sensitivity and Classroom Team Projects: An Empirical Investigation," *Proceedings of the 43rd ACM Technical Symposium on Computer Science Education* (2012): 403–408.

319 **as well as in *online collaborative settings*:** David Engel et al., "Reading the Mind in the Eyes or Reading Between the Lines? Theory of Mind Predicts Collective Intelligence Equally Well Online and Face-to-Face," *PloS One* 9, no. 12 (2014): e115212.

INDEX

ABOUT THE AUTHOR

Chantel Prat is a professor at the University of Washington with appointments in the departments of Psychology, Neuroscience, and Linguistics; with affiliations at the Institute for Learning and Brain Sciences, the Center for Neurotechnology, and the Institute for Neuroengineering. She is a recipient of the Tom Trabasso Young Investigator Award from the Society for Text & Discourse and a Pathway to Independence Award from the National Institutes of Health. Prat also speaks internationally at events like the World Science Festival. She is featured in the documentary film *I Am Human*. Her studies have been profiled in media ranging from *Scientific American, Psychology Today,* and Science Daily to *Rolling Stone, Popular Mechanics, Pacific Standard, Travel + Leisure,* and National Public Radio.